Exploring the Value of Electricity

Exploring the Value of Electricity

Clark W. Gellings, P.E.

Published 2020 by River Publishers
River Publishers
Alsbjergvej 10, 9260 Gistrup, Denmark
www.riverpublishers.com

Distributed exclusively by Routledge
4 Park Square, Milton Park, Abingdon, Oxon OX14 4RN
605 Third Avenue, New York, NY 10017, USA

First issued in paperback 2023

Library of Congress Cataloging-in-Publication Data

Gellings, Clark W.
Exploring the value of electricity/ Clark W. Gellings, P.E. -- 1 Edition.
pages cm
Includes bibliographical references and index.
0-88173-748-8 (alk. paper) -- ISBN 978-8-7702-2326-3 (electronic) -- ISBN 978-1-4987-4431-7 (Taylor & Francis distribution : alk. paper) 1. Energy consumption. 2. Energy conservation. I. Title.

HD9502.A2G435 2015
333.793'2--dc23

2015015303

Exploring the Value of Electricity/ Clark W. Gellings.
First published by Fairmont Press in 2015.

Routledge is an imprint of the Taylor & Francis Group, an informa business

Publisher's Note
The publisher has gone to great lengths to ensure the quality of this reprint but points out that some imperfections in the original copies may be apparent.

ISBN 13: 978-87-7022-934-0 (pbk)
ISBN 13: 978-1-4987-4431-7 (hbk)
ISBN 13: 978-8-7702-2326-3 (online)
ISBN 13: 978-1-0031-5195-1 (ebook master)

Contents

Chapter 1

Introduction—Electricity is Valuable!

Few dispute the fact that electricity is valuable. Certainly electricity provides services, sustains and improves the quality of life, provides warmth, illumination, transportation, medical diagnostics, and motive power, facilitates product development and use, and engenders technology. But electricity's value is not limited to these categories nor can its value be demonstrated by narrow examples of its use. Electricity empowers society and its synergistic effects offer benefits which have hardly been identified, let alone explored, evaluated or documented.

SUMMARY

There are numerous viewpoints regarding electricity's value as are elucidated throughout this book.

Today electricity is literally woven into the fabric of modern society. To do without it is inconceivable, yet to estimate its value suggests there is a model in which we can compare life with and without it.

Electricity is taken for granted, and why shouldn't it be? It has become an essential service, like water, telephone, streets and bridges, the value of which are not easy to estimate.

One occasionally cited measure is to assume that the nation's gross domestic product (U.S. GDP) is entirely dependent on electricity. That would suggest that the value of electricity is $17.6 trillion/year or the equivalent of $55 thousand per person—multiple orders of magnitude greater than anyone's electric bill.

Based on an estimate of U.S. population as of July 1, 2014, of 322.6 million, just looking at electricity's role in medicine and its ability to power communications and sensors related to prolonging life rolls up to $17.2 trillion/year.

Using outage as a measure implies even greater value, since just one large industrial concern can lose more than $140 million/day for a prolonged outage. One data center reported losses of nearly $2.0 million for just one day. While each major U.S. storm costs society between $150 and $240 billion/year.

Using customer surveys of blackouts shows an enormous range of impact from $.17/kWh to $366.00/kWh, depending on customer type, size and duration of outage.

In this book we discuss each of these and more.

THE VALUE OF ELECTRICITY STORYLINE

Electricity is a natural phenomenon—one of the basic forces of nature. It was *not* invented by humankind, it was discovered by humankind. It would be foolish to speculate life without electricity or electric phenomena. In fact without electricity, there would be no life—all living creatures rely on electricity to send impulses to and from nerves and muscles. However, it is possible to speculate life without a modern electric system—a world that would be void of many, if not all, of the modern conveniences society now enjoys. And, it is possible to speculate on the attributes of electricity's value. Both are attempted in this book.

The value of electricity which society acknowledges can be exemplified by an engraving chiseled in the building which is the Union [railroad] Station in Washington, DC. The engraving honors the progress of railroading and sits high above the West Entrance flanked by Greek demigods Thales (representing electricity) and Themis (representing freedom or justice). It reads:

> "FIRE—GREATEST OF DISCOVERIES ENABLING MAN TO LIVE IN VARIOUS CLIMATES, USE MANY FOODS—AND COMPEL THE FORCES OF NATURE TO DO HIS WORK.

Figure 1-1. Illustrates two worlds—one with Electricity and one without

> ELECTRICITY—CARRIER OF LIGHT AND POWER, DEVOURER OF TIME AND SPACE—BEARER OF HUMAN SPEECH OVER LAND AND SEA, GREATEST SERVANT OF MAN—ITSELF UNKNOWN."
>
> THOU HAS PUT ALL THINGS UNDER HIS FEET

It stands as a reminder—not just of railroading—but of electricity as the "greatest servant of man."

Electricity Drives Human Functions

Electricity is key to the function of the human body—it allows messages to be sent from one point to another, controls the beating heart, triggers muscle movement, and facilitates memory. The body generates electricity by leveraging the slight imbalance between the potassium and sodium ions inside and outside the body's cells. At rest, the body's cells have more negatively charged potassium ions inside the cells than positively charged sodium ions. When the body needs to send a message, a gate opens and ions flow generating an electrical impulse. Without electricity, the human body would not function. Scientists first began to realize the role electricity played in bodily functions by experimenting with animals. For example, in 1786, Galvani made this discovery by stimulating responses in frog legs with electricity. (Simon 2004)

One example of the dependency of the human body on electricity is evidenced by observing the phenomenon of epilepsy. As highlighted in the book, *Electricity* by Ray Robinson, an epileptic seizure actually is an electrical phenomenon. The story describes a woman who lives with epilepsy, where uncontrollable surges of electricity leave her in a constant state of edginess. She sees the world in terms of angles: "you look at every surface, you weigh up every corner, and you think of your head slamming into it. Prickly, up-front honest, and down-to-earth practical, Lily has learned to make do, to make the most of things, and to look after—and out for—herself." (Robinson 1971)

Robinson describes her ordeal as the following:

"People will be stood over me, faces looming out of the dark after-fuzz, not knowing what to do. The old man pulling the dog off from licking me. And I'm like if I don't wet myself, then maybe Ridge Racer won't mind me being late. Maybe he's late too. Or maybe I can rush home and change anyway, or just stick my knickers in my bag. Dry the back of my skirt under the blow-dryer in the ladies, spray some perfume on, right as rain."

What makes your brain do that? Electricity.

Electricity can actually help prevent the failure of the human body's electrical system. Many young people die of the failure of the heart's electrical system known as sudden cardiac arrest or SCA (Winkley 2013). SCA is not a heart attack. It occurs when the electrical system in the heart stops working rather than when a blockage disrupts blood flow. Thousands of teenagers die every year from sudden cardiac arrest. Many show no warning signs and have no family history of heart defects. However, a risk for sudden cardiac arrest can be identified with a noninvasive test called an electrocardiogram or EKG. An EKG tests the heart's electrical system and can alert physicians to warning signs.

It has been speculated that chemical energy from the blood can be converted to electrical energy. Panasonic's Nanotechnology Research Laboratory has experimented with enzymes that can strip blood glucose of its electrons to create a charge. (smh.com.au/articles/2003)

THE VALUE OF ELECTRICITY

Living Without Electricity

Papua New Guinea—The morning sun peaked in through the slats in the east wall of Kama's family hut. The family stirred. It was early morning in an isolated village high in the mountains of Papua New Guinea (PNG). PNG occupies the eastern end of the island of New Guinea. The 6.5 million indigenous inhabitants of New Guinea comprise one of the most heterogeneous populations in the world; PNG has several thousand separate communities, most with only a few hundred people; divided by language, customs, and tradition, some of these communities have engaged in low-scale tribal conflict with their neighbors for centuries. Many of the villages are without electricity. As a result there is no modern communication, no internet, no telephone, no radio, no TV and nothing to display a standard of living and acceptable human behavior beyond what they have known forever. There are a few radio stations on the island that broadcast in their native language, Tok Pisin, a creole language, even if they had a radio. A few of the villagers have been exposed to modern society and the potential virtues of energy sources like electricity, as evidenced by the thread bare tee shirts some of them sport. But there is little recognition as to how an energy form like electricity could possibly change their lives in so many ways.

PNG ranks 129th in the world in per capita electricity production with only 700 MW total generation on the island. There are virtually no jobs to go to, no money to earn and no coins or dollar bills. In fact the only currency is livestock—pigs and chickens. Which are used for rare meals of meat and only slaughtered for special feasts. As currency, pigs are frequently use to pay a dowry for a new bride. Pigs are one of the few symbols of wealth. The entire focus of Kama and his wife, Loujaya's days are to find food to support themselves and their four children. Their diet is referred to as a Paleolithic diet and consists of mainly tubers (yams, sweet potatoes and taro), fruit, fish and coconut. Bread, made from dough pounded from the sago palm supplements their diet. The tribes are basically farmers, but they do hunt the occasional bird and those who live near the coast do hike down to the beach most days to fish. According to the press reports, they have among the lowest heart circulatory problems in the world. There is no medical care except what they can provide themselves using ancient traditions of healing. And if modern vaccines were transported here through the jungle—there is no refrigeration to preserve them.

Kama arose slowly that morning—his muscles ached as he had been out all day before clearing a patch in the jungle to grow a new crop of tubers. To clear such a patch, he had to first trek around the jungle for some time to find a prospective clearing—one with not too many large trees, then use hand tools to laboriously remove the smaller brush and some of the trees. Still to be done in the days ahead, holes need to be dug where shoots from a recent harvest of mature tubers are then planted. After 12 hours in the hot sun, Kama was tired and sore. Sure footed, jungle hardened feet carry these people on their treks around the country side; there are no athletic shoes stitched together by electric powered sewing machines.

Tubers preserve well, but not so fish and other meat. Without electricity they have learned to harvest only what they need. Sufficient tubers for that day's meal, or Sago palm meat to pound into a dough and bake over an open fire are harvested sparingly along with Kakau, baked sweet potato. Loujaya has decided to make Kakoda Fish today, a fish dish cooked with coconut milk and lime. She sets out on a long hike with two oldest daughters to the ocean, carrying nets hand made from vines to "net fish"—casting nets in the pools along the seashore teaming with fish.

As they finish the evening meal and the campfire provides lingering light, there are stories to be told of the day's harvest, of the toil in the fields and of the lore of conquests of day gone by.

But there are no books to read, no math homework for the children. They can't read nor write, there are no schools in this or any of the other villages in the region, and no sources of light aside from the fire glow by which to read them. There are no expectations that their children will live any differently than they do now. And meanwhile, somewhere in the western world, teachers are debating how to engage more students in STEM (Science, Technology, Engineering and Math) programs.

Near Johannesburg, South Africa—It was my first visit to The Republic of South Africa and Eskom, the major electric utility and one of the most functional South African institutions was my host. That day we were to visit a village—a collection of dwellings in the suburbs of Johannesburg, to learn about the impacts of electrification. These dwellings consisted of structures referred to as "informal housing" and varied from concrete block, one room building to various wood frame structures or sheet metal enclosures and even shacks with plywood sides supported by wood poles. Then in the 1990s, over one million South Africans were without electricity and I was witnessing the beginning of an aggressive program, which still ongoing, to electrify them. The village we were visiting was in the process of being electrified with about ½ of the homes now with power.

The Eskom team of two distribution engineers, Ray and Chris, were dressed informally and did not wear any Eskom insignias. They carried no weapons and wore no bullet proof vests. I did not realize the significance of their lack of personal protective gear until the next day when I was escorted by government representative on an official visit. They dressed me in a vest and stood by my side with automatic weapons at the ready. The Eskom team had arranged a meeting with community leader, a Pentecostal Bishop who had gathered a group of about six parishioners in his church to discuss electrification. We parked some distance from the church where we met him. As we walked on the dirt road toward the community, townspeople would approach, and as they did he would put his arm on my shoulder—I later realized he was signaling that I was a friend and not to be harmed.

We were passed by a woman with a wheelbarrow. In the wheelbarrow were three lead-acid automotive batteries. She was returning from the welding shop some distance away where the batteries had been recharged. The welding shop was in the sector which had already been electrified and the woman's home was in a sector which had not. The batteries were used primarily for lighting and a radio and an occasional direct current (DC) TV. There were already signs of economic activity supported by electricity which the Eskom team pointed out along the way. In addition to the welding shop, there were crude signs indicating "Panel Beaters" (auto body repair), a hand drawn sign "Beauty Parlor" and a few homes which had refrigeration and for a nominal fee would host local community citizens to a beer. In the more established areas, some convenience stores had taken root with commercial refrigerators where cold drinks and refrigerated food could be had.

The church was a rather informal structure a main supports consisting of medium sized tree trunks with a combination of plywood, sheet metal and fiberglass panels scavenged from somewhere as walls and a roof. Only half of the roof was complete and we sat in the area where it was complete. My meeting with the townspeople was to have been centered primarily on safety. It reflected a concern that these South Africans did not adequately yet understand that electricity could kill. In townships where power had already been provided, there had been deaths. Some of these deaths occurred from attempts to divert electricity or steal it. Eskom had invented an ingenious service wire which delivered 220 volt power to buildings. It consisted of an insulated copper core surrounded by an aluminum sheath which acted as the neutral which was again covered by insulation. This configuration made it harder to break into the core without shorting out the circuit. But it made theft a more dangerous undertaking. We discussed theft and the dangers of handling electrical wiring as had been illustrated nicely in a then recent set of published Eskom illustrations.

One of the subjects they raised was regarding what they

called "boycott meters." I must confess that at the time I did not have a clue as to what they were speaking about. Having been introduced as a world expert, I hesitated to demonstrate my ignorance in this venue. The newly electrified communities were being outfitted with prepayment meters. Meters which required a code which you obtained by purchasing a quantity of electricity. You inputted this code and the meter would display and track your usage of electricity. The term boycott meter referred to the traditional kilowatt-hour meter. The term originated from an organized boycott of electricity payment which sprang up in the oldest South African Township, another suburb of Johannesburg called SOWETO, or South West Township. The citizens of SOWETO had stopped paying their electric bills and stopped letting meter readers in and blocking access to any Eskom workers who might try to interrupt their supply. The result is they were getting free electricity. The SOWETO movement started as part of the anti-apartheid protests. My parishioners were asking "I don't understand why we couldn't have "boycott meters" as a way to ask "Why couldn't we have free electricity."

The new electrification was bringing them refrigeration, lighting, TV, radio, a cook top and an occasional 'geezer' (hot water heater). One of the contractors which electrified the 500th home in the township awarded the homeowner with an electric stove. She was really proud of that stove and decorated it with some nick knacks. I asked what she did with the nick knacks when she uses the stove and she responded that she only used the stove once and the meter "spun so fast" she turned it off and will not use it again. Figure 1-2 depicts scenes from a South African Township.

Overall the phenomenal impact of electrification to these South African communities was obvious and they were so proud of the developments. As in most parts of the world, the impact on the standard of living and on economic development was nearly instantaneous.

Brazil, India and Indonesia—As I travelled through other parts of the Third World, from Manaus in the Amazon region of Brazil,

Figure 1-2. Scenes from a recently electrified South African Township (Photos by the author)

to rural India north of Bangalore and to a fishing villages in Indonesia, much of the story remained the same with minor variations. In most of these other regions, the indigenous people were aware of electricity and of modern conveniences. While they wanted to have electricity, it was either simply not available or was not affordable. In many cases, even when it is available, the power quality and overall reliability is poor. As the networks of electrification expand and form more reliable webs, these attributes of service will continue to improve. In these other countries, aside from a greater awareness of electricity and modern conveniences, three other elements stood out.

First the availability of medicines was facilitated by the use of small refrigerators in remote communities, usually powered by photovoltaic devices and batteries. These distributed generation and storage systems paved the way to cost reductions which would later allow other distributed installations. Secondly, the proliferation of cell phones and mechanisms to recharge them. Within the last ten years or so, simple, non-smart cell phones began to appear in remote areas throughout most of the remote regions of the world. With the phone came the issue of recharging them and several schemes evolved. In some cases indigenous people could rent a charging outlet for an exorbitant fee for a period of time. In others, batteries could be exchanged. The most innovative new application is the development of a device using a thermocouple which is inserted into the cooking fire or hearth and generates enough electricity from the conversion of heat to electricity so as to charge the cell phone. Third, many remote communities have begun to have exposure to television and the internet. These were generally accessible at some community center. In these cases distributed generation systems were often used. Many such systems were sponsored by governments, Non-Government Organizations (NGOs) and even private developers who began to see a profitable new business in selling electricity to communities. The combination of television and the internet catalyze education and made even remote schools more complete and capable. And as the importance of education was accepted more widely, the demand for electric lighting increased. Light at night meant more time to study and read.

In all these countries there were weaknesses in essential community services. While in South Africa I was told by local officials about a new water "tap" (source of fresh water) that was installed recently just across the highway from a township. Unfortunately, seven citizens from the township had already been killed in the last few months trying to dash across the highway to access the tap. It had not occurred to them to relocate the tap to the opposite side of the road. Many communities in these Third World setting were beginning to embrace education, however access to books, computers and the internet—as well as power to energize them were only sparsely available. And as I began to lead an effort for EPRI to bring advanced technologies for essential community services to South Africa, the first technology I was encouraged to research were latrine pit liners. These liners were laid on the surface of road side pits were people urinated. The liner minimized flies and insects letting the urine sift through and soak into the earth.

A common problem in all communities is air pollution—particularly indoors. While this varies depending on the climate, with colder climates having the greatest problem, the combustion of wood, dung or dried peat moss indoors in a poorly ventilated building to illuminate, cook or heat causes a substantial health risk to these citizens.

If you never saw electric appliances and devices—or witnesses the benefits that they provide in enhancing the quality of life—then you would never comprehend how desirable it could be. As electricity begins to appear in areas of the world where it was not generally available, then, it becomes a convenience and gradually progresses to an essential community service.

The Human Development Index

Over one billion people in the world depend on wood, animal dung, kerosene or similar, low grade fuels. In addition, they utilize these fuels in very inefficient appliances and devices which deliver energy services to the poor such as lighting, heating and cooking. Comparing access to electricity in nations which have electricity with nations who have limited access amount its population gives insight to its value. As shown in Figure 1-3 just a few kilowatt-hours per capita of annual consumption yields an enormous increase in human development, as depicted by the Human Development Index (HDI).

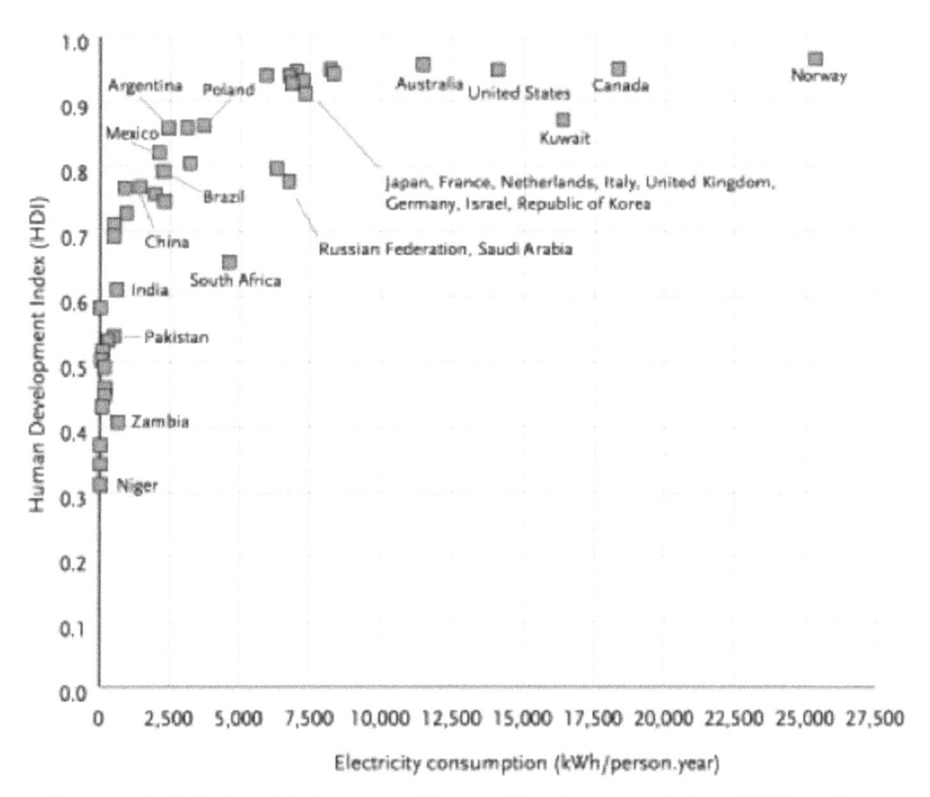

Figure 1-3. Relationship between Human Development Index (HDI) and per capita Annual Electricity Consumption (United Nations Development Programme, 2006 and Schnitzer, 2014.)

The HDI was developed by the United Nations Development Programme to emphasize that people and their capabilities should be the ultimate criteria for assessing the development of a country, not economic growth alone. The HDI is a summary measure of average achievement in key dimensions of human development: A long and healthy life; being knowledgeable; and having a decent standard of living. The HDI is the geometric means of normalized indices for each of the three dimensions. (http//hdr.undp.org)

Among the HDI drivers are the availability of artificial illumination. Figure 1-3 helps support this:

> As observed by photographer, Peter DiCampo in 2013: "While living and working as a volunteer in remote Northern Ghana, I realized how deeply the lack of electricity affected the lives of my neighbors. It impeded their progress in the sectors of health, education, gender equality, agriculture, and virtually every aspect of development. And, of course, there's the lack of light. The situation in Northern Ghana is not uncommon: in fact, many places the implications of living without light are far more dire." (http://www.lifewithoutlights.com).

A blog (ErgDev) on February 8, 2010 posted the results of a survey whose respondents were asked to characterize the most important benefits of electrification. The study acknowledged how difficult it is to measure the value of electricity in many countries because access to it is virtually universal, and prices and connection costs are often subsidized or set by regulatory agencies." The results indicated that the most important benefits of electrification were:

- Lighting (artificial illumination)—42%
- Productive Uses—28%
- Cooling or Heating (space conditioning)—0%
- Health Clinics or Other Public Uses—7%
- Children's Education—7%
- Television or Radio—14%

THE BASIC VALUE OF ELECTRICITY

More effectively than any other energy form, electricity provides light, heat, comfort, mechanical work, powers the digital age and enhances the quality of life. Electricity can meet society's need in an affordable, reliable, economical and environmentally sensitive manner.*

- Electricity can fully leverage the use of zero emission renewable energy resources.
- Electricity generation can effectively be based on photovoltaics, concentrating solar power, wind, nuclear power, biopower (electricity from sustainable biomass), hydro and kinetic (run-of-river, tidal and wave) and geothermal energy resources which have zero emissions.
- Electricity can provide access to the entire electromagnetic spectrum enabling infrared, x-rays, ultraviolet rays, radio, frequencies, microwaves, ultrasound, high-frequency AC for artificial illumination, Lorenz forces to turn motors, and other electrical phenomena which leverage energy input by up to several magnitudes.
- Can be precisely controlled and directed with near unlimited quantity with precision.
- Enables the digital economy and communications and entertainment.

Electricity is a uniquely valuable versatile and efficient form of energy, offering unmatched precision and control in application. Electricity offers society more than just improved energy efficiency. It also has greater "form value" than any other energy source: form value affords technical innovation with enormous potential for economic efficiency. Form value encompasses three dimensions: technical, economic, and resource uses.

*This section borrows heavily from *Saving Energy & Reducing CO_2 Emissions with Electricity*, C.W. Gellings, The Fairmont Press, Inc., Lilburn GA, 2011

Gateway to the Electromagnetic Spectrum

The technical dimensions enables access to the entire electromagnetic spectrum. The electromagnetic spectrum consists of all forms of electromagnetic radiation, each corresponding to a different section, or band, of the spectrum. For example, one band includes radiation that our eardrums use to interpret sound. While another band, visible light, consists of radiation our eyes use to "see" light.

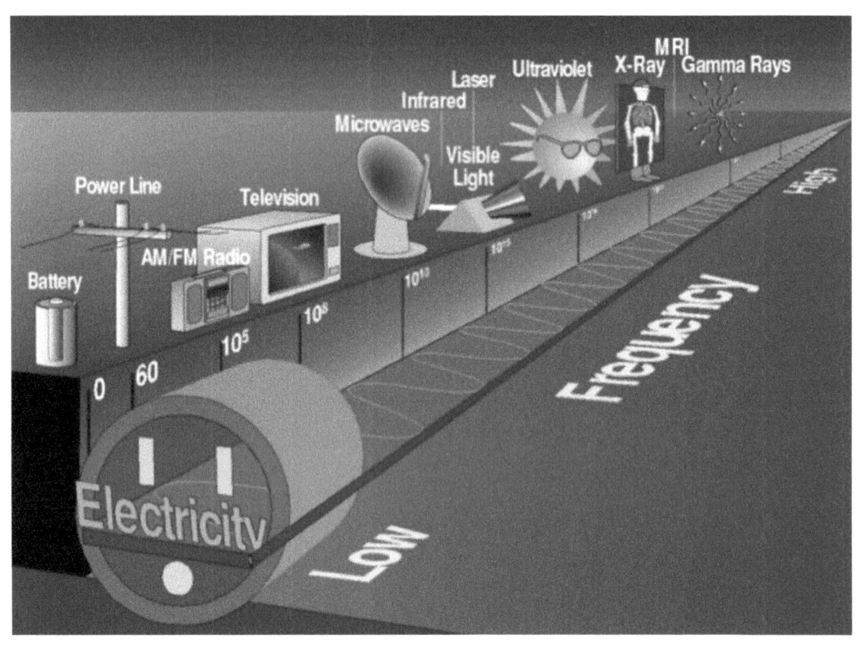

Figure 1-4. Gateway to the Electromagnetic Spectrum (C.W. Gellings, Electric Power Research Institute)

Electricity, or electromagnetic energy, is the only energy from which can provide a "gateway" to the electromagnetic spectrum. The electromagnetic spectrum embodies the range of all frequencies of electromagnetic radiation. It extends from frequencies below those used for electric power distributed (50 to 60 hertz (Hz)) to those used for AM/FM radio, television, microwave ovens, radar, infrared visible light, ultraviolet light, x-rays and gamma rays.

The economic dimension is where electric-powered end-use appliances offer significant advantages when compared with conventional (e.g., thermal) counterparts such as natural gas or oil including reduced energy use, increased productivity, and improved product quality, compactness, and environmental cleanliness. The dimension of resource use is where electricity can be produced from a range of fossil fuels, nuclear, and renewables. In addition, electricity has the following key attributes:

Input Energy Density—in combustion process using chemical fuels (e.g., oil and gas), the maximum achievable temperature is thermodynamically limited to the "adiabatic flame temperature," a practical limit of about 3000°F for fossil fuels burned in air. When heating material electrically, there is no inherent thermodynamic limit on the temperature.

Volumetric Energy Deposition—Electrothermal phenomena are volumetric (i.e. generating heat within the material itself). When using fossil fuels to heat material, heat is usually imposed at the surface by radiation and convection. This method is inherently slow and inefficient. With induction heating, electrical energy is deposited directly within the material; thus, processing time can be reduced to several minutes or less. Volumetric heating can also be used to dry moist materials with microwave or radio frequency radiation.

Controllability—Electricity is an "orderly' form of energy, in contrast to thermal energy which is random in nature. Electrical processes can be controlled much more precisely than thermal process.

Flexibility of Raw Material Base—The high-energy intensity and precise control offered by electro technologies permit a greater degree of flexibility with regard to raw material resources than do fossil-fueled processes.

Product Quality and Yield—Electrical processes are superior in providing product quality and yield.

Resource Use—Resource use attributes of electric processes include use of renewable resources, flexibility of fuel supply; domestic resource balance of payments; national security; environmental; and energy consumption. Electricity offers the opportunity to use a number of low or zero carbon-emission sources of electric power production.

Environmental—Electric processes and systems are undisputedly the most environmentally benign at the point of end use.

Energy Consumption—Primary energy consumption is almost always lower for electrical processes than with conventionally fueled systems.

HOW DID MODERN POWER SYSTEMS EVOLVE?

Electricity has become so ubiquitous and is so valuable that it is now difficult to make any reasonable estimate of its value, but—what if the modern electricity infrastructure which laces through the western world, never evolved as it has. How different the world would be and how extensive the value that it derives would have been lost.

The Telegraph Leads the Way

The development of primitive electric power systems followed the development of the telegraph. In 1828, the first telegraph in the USA was invented by Harrison Dyar who sent electrical sparks through chemically treated paper tape to burn dots and dashes. (Bells, 2014)

In 1825, British inventor William Sturgeon invented the electromagnet. The electromagnet laid the foundations for large-scale evolution in electronic communications. In 1830, Joseph Henry, an American, demonstrated the potential of William Sturgeon's electromagnet for long distance communication by sending an electronic current over one mile of wire to activate an electromagnet which caused a bell to strike. In 1837, British physicists, William Cooke and Charles Wheatstone patented the Cooke and Wheatstone telegraph using the same principle of electromagnetism.

However, it was Samuel Morse that successfully exploited the electromagnet and bettered Joseph Henry's invention. Morse invented a telegraph system that was a practical and commercial success. Samuel Morse proved that signals could be transmitted by wire. He used pulses of current to deflect an electromagnet, which moved a marker to produce written codes on a strip of paper resulting in the invention of Morse code. Later, the device was modified to emboss the paper with dots and dashes.

In 1838 Congress funded $30,000 to construct an experimental telegraph line from Washington to Baltimore based on Morse's inventions.

Samuel Morse and his associates obtained private funds to extend their line to Philadelphia and New York. Small telegraph companies, meanwhile began functioning in the East, South and Midwest. Dispatching trains by telegraph stared in 1851, the same year Western Union began business. Western Union built its first transcontinental telegraph line in 1861, mainly along railroad rights-of-way.

Until 1877, all rapid long distance communication depended upon the telegraph. That year, a rival technology developed that would again change the face of communication—the telephone. By 1879, patent litigation between Western Union and the infant telephone system ended in an agreement that largely separated the two services.

In some measure, it was the demonstration that wires could be strung long distances in telegraph systems mounted on poles along roads and railroad tracks that let to acceptance of early electric systems. (Bells, 2014)

The Evolution of Power Systems

One of the first challenges which both Edison and Westinghouse faced in the operation of either the first direct current (DC) power systems or alternating current (AC) power systems was to enable reliable operation through control such that in any instant the total generation in a power system was "balanced against" total load. Balancing ensures the generation which is running at any point in time be equal to the total load or demand for electricity at that same moment. When generation is not balanced with load, the system becomes unstable and can collapse. In the 1800s when the first of the Pearl Street Generators was placed in service, this balancing act was a relatively simple endeavor primarily done through control systems in generators.

As power systems evolved, they became larger, more complex, and increasingly more difficult to control. Balancing multiple generators with a network of loads located throughout a city or across town and into the countryside could not be facilitated by generation control alone. By the mid-1950s, some of the first Supervisory Control and Data Acquisition (SCADA) systems were being deployed within power delivery systems. This development and the others which have led us to the modern power system which we now enjoy, have allowed us to benefit from the value which electricity provides.

Electricity's Timeline Interrupted?

What if electricity's discoveries began in 600 BC and continued through 1886—but were then interrupted? What would that time line look like? This assumes that electricity was discovered and some of its attributes exploited, but that the electricity infrastructure including the utilities which serve society never matured.

The Discovery of Electric Phenomenon

600 BC Thales experiments with static electricity.

1660s Otto von Guerick invents a static electricity machine which mimics the generation of static electricity in nature.

1745 Pieter van Musschenbroek and Ewald Georg von Kleist collaborate on the invention of what is now called a Leyden jar—the first capacitor.

1752 Benjamin Franklin demonstrates that lightning is electricity (and amazingly he doesn't die in the process).

1790s Alessandro Volta creates the first battery (the concept of potential electrical difference—the volt, is named after him).

1820 Hans Oersted finds a connection between electricity and magnetism—hence the ability to generate electricity with motion, wires and magnetism.

1831 Michael Faraday conceives electromagnetic induction.

Modern Electric Inventions

1879 Thomas Edison invents the first practical light bulb.

1881 Electricity evolved in the era of the expansion of railroads. In 1881-1882, railroads had laid an astounding 10,000 miles of track creating an atmosphere that innovators could do most anything. (Jones 2004) As the railroads expanded, so did the telegraph.

1882 Thomas Edison constructs and commissions the first functional power system.

1886 George Westinghouse, Jr. constructs and commissions the first alternating current power systems based on inventions by Nicola Tesla.

Electricity Timeline Interrupted—What if at this point in this timeline the development of electric power systems is halted—or never evolves! Electricity systems are demonstrated but are never a business success.

Revolution

A society without electricity was dramatically depicted by a recent TV series.

"Revolution" aired from 2012-2014 and was based on earth fifteen years after a blackout where a group of revolutionaries seeks to drive out an occupying force posing as the U.S. government.

The story line reinforces that our entire way of life depends on electricity and speculates as to what would happen if it just stopped working. It alleges that, one day, like a switch turned off, the world is suddenly thrust back into the dark ages. Planes fall from the sky, hospitals shut down, and communication is impossible. And without any modern technology, who can tell citizens why? Now, 15 years later, life is back to what it once was long before the industrial revolution: families living in quiet cul-de-sacs, and when the sun goes down lanterns and candles are lit. the script suggests that without electricity, life is slower and sweeter. Or is it? On the fringes of small farming communities, danger lurks. And a young woman's life is dramatically changed when a local militia arrives and kills her father, who mysteriously—and unbeknownst to her—had something to do with the blackout. This brutal encounter sets her and two unlikely companions off on a daring coming-of-age journey to find answers about the past in the hopes of reclaiming the future.

In "Revolution," they lived in an electric world. They relied on electricity for everything. When the power went out, everything stopped working and they weren't prepared. In the show, fear and confusion lead to panic; the lucky ones made it out of the cities; the government collapsed; and militias took over, controlling the food supply and stockpiling weapons. (www.imdb.com).

In another recent movie where apes living without electricity acquire human intelligence and face direct conflict with the humans who managed to survive near destruction of the planet, a human remarks:

"You know the scary thing there: they [apes] don't need power—lights, heat, nothing, that's their advantage, that's what makes them stronger," (Source: "Dawn of the Planet of the Apes" movie trailer—human survivor commenting on the apes who had taken over the works.)

This book is about a quest to understand the value of electricity. In part it speculates as to what life would be like if this quest were doomed. Electricity has become so intertwined, so essential to the way of life of the peoples of the western world that it can be equated to the value of air, clean water, bridges, and telephones. It is no longer possible to use traditional energy economics to calculate electricity supply and demand curves and to integrate them to estimate the value.

A renowned scientist and a colleague at EPRI, Victor Niemeyer, suggested that perhaps one way to describe the value of electricity was to envision the world without it. What if Franklin, Tesla, Faraday, Edison, Steinmetz, and the other great pioneers had not made their discoveries and that the mystery of electricity eluded us even into the 21st century—what would the world actually be like? Dr. Niemeyer suggested that the plot for such a story might follow the kind of storyline used by several authors to describe the world as if Germany during Hitler's time had won World War II. [See for example "What if Hitler Won the War? (Where would we be Today)" by Philip Moore.]

What if the early stages of society's introduction to electricity went terribly wrong? What would a "modern" household be like without electricity? And what would an average day be like without electricity in the 21st century?

Chapter 2

What if There Were No Electricity?

One Wiki entry (www.wiki.com) addressed the question: "What would happen if there was no electricity?" Wiki Answers offers two responses: (1) that the universe would not exist since if electricity did not exist and since it is comprised of electron movement and therefore electrons—then electrons would not exist and therefore matter as we know it would not exist; and (2) that humans never learned to use electricity. The second response infers that society never moved beyond the technologies of the mid-1880s where appliances were human-powered, lighting, heating and cooking was provided by burning fossil fuels of sorts, and machines were water or steam-powered and reading—albeit limited without light—provided most of the personal entertainment instead of electronics.

WHAT IF HUMANS DID NOT KNOW HOW TO USE ELECTRICITY

Wikipedia speculated as to what changes would have occurred in society if electricity were not available. They are as follows:

- There would be no radio, TV or computers
- There would be no microwave ovens
- Cars would be diesel-powered only [with pneumatic starters]
- There would be no [electrically powered] escalators or elevators
- Only mechanical handheld calculators would be available

- Only mechanical typewriters would be available
- People would more likely go to bed at dark

In the early 1940s a psychologist, A. Maslow, outlined what has been referred to as the theory of hierarchy. (Maslow 1943). In it he outlined five elements which summarized human motivation: physiological; safety; love/belonging; esteem; and self-actualization. One way to speculate on life without electricity is to align Malow's theory with the conveniences which electricity does provide. Figure 2-1 illustrates this relationship.

The Dismal Science

The philosopher, Thomas Malthus, published in 1798 in "An Essay on the Principle of Population" a treatise on the relationship between technology and population. In his writings, he chose the relationship between world population growth and agricultural production. Agricultural production has historically increased because of technology development and the discovery and development of electricity. Electricity is, in fact, a technology development achievement of its own.

Malthus theorized that population growth was the ultimate measure of the growth of humanity and provided evidence that population growth from the early 1900s forward was triggered by the ability to enhance food production. The same theory could be applied to electricity and its adjacent technologies. That is, the growth in population, and perhaps the quality of life itself, is because of the invention, and expansion—the proliferation of electricity.

As a result, Malthus has been credited with inventing "The Dismal Science." His theories have been credited with establishing the relationship between technology and population growth—and the concept that population growth is limited by technology. In a brief written for "Policy Analysis" in 2012, Indur Goklany postulates that transforming human nature from a dependency on nature to one that also embraces technology expands every indicator of well-being "...such as levels of hunger, infant

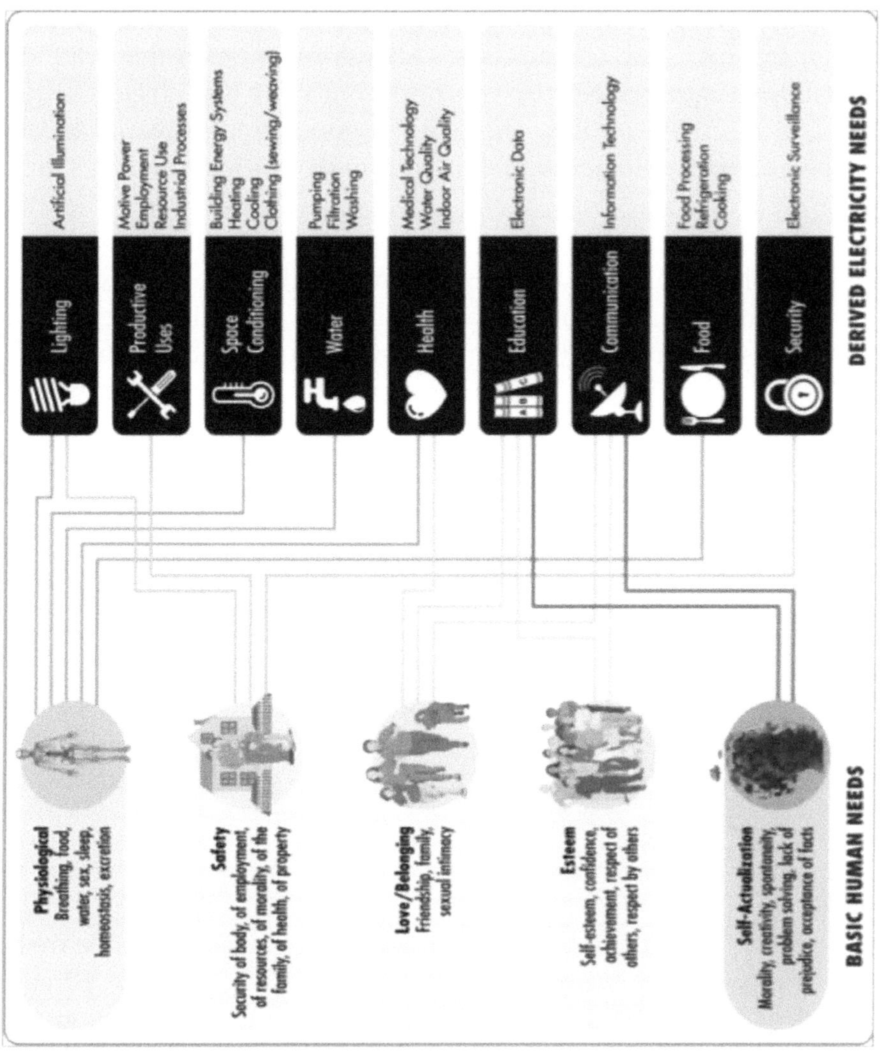

Figure 2-1. Illustrates the relationship between Maslow's theory and electricity's attributes

mortality, life expectancy, education, economic freedom, and child labor improves as income rises." Therefore, one possible way to establish a value for electricity is to assume there was no electricity and to speculate on how the world's population(s) would have evolved. Some elements of this approach are included in later chapters.

Other indirect references to electricity and this technology relationship (Ellperin 2012) support this theory, for example suggesting that technology can reduce the mortality of citizens by the use of air conditioning. His study suggest that mortality effect of an extremely hot day declined by about 80% between the periods 1900-1959 and 1960-2004. The study found that days with temperatures exceeding 90°F were responsible for about 600 premature fatalities annually in the 1960-2004 period, compared to the approximately 3,600 premature fatalities that would have occurred if the temperature-mortality relationship from before 1960 still prevailed. It was during this period that air conditioning was widely adapted.

Value of Electricity:
Hypothesis: Air conditioning saves lives
Assumptions: 91 million homes save 3000 lives/per year @ 3833 kWh/home. Life worth $5.5 Million (EPA 2004)
Value of Electricity: $.054/kWh

It has been 130 years since Thomas Edison energized the first electricity supply system on September 4, 1882, at Pearl Street in New York. Networks of electrical systems have evolved to electrify much of the world which in 2003, the National Academy of Engineering identified electrification as the top engineering achievement of the 20th century.

But perhaps it was not to be!

On September 16, 1878, the *New York Sun* proclaimed that Edison's newest marvel will send cheap light, heat and power by electricity. And for a short time, the public believed that electricity was destined to be the agent of change for society.

WHAT IF ELECTRICITY FAILED?

Electricity could have failed to develop because of a related combination of public opinion and the inability to develop an effective business model. Electricity could have been simply viewed as too dangerous, too expensive, and of not providing sufficient value to overcome those considerations.

In the 1880s, electrical components were very costly, newspapers frequently reported electrical fires and electrocutions, magazines offered long lists of cautions for those who were adventurous enough to install electricity. (Simon 2004) One author exclaimed that "What electricity generated most pervasively was anxiety." (Simon 2004)

The thesis that electricity never evolved is not so far-fetched: 30 years after Edison established the first electric distribution utility in 1879, barely 10% of American homes had electric energy service. Numerous articles extolled the hyperbolic claims of electrical companies which enhanced the public's concerns of this new, untrusted technology. Accidental electrocutions and electrical shorts or flashes made news. (Simon 2004) Articles warned about the adverse effects of electricity including blindness, all pointed to the alignment of electricity to the devastating power of lightning.

One exception is that for a time, the public accepted electrotherapy—widely used in 19th century medicine, but still other concerns grew and festered.

A Fear of Electricity

Some people do have a fear of electricity—called electrophobia, it prevents people from going outdoors during thunderstorms or from operating appliances. These people may even be afraid to be in the same room with an electrical appliance. (Ezine articles) Phobias arise from a combination of external events such as a traumatic event possibly coupled with internal predispositions based on heredity or genetics. Symptoms vary but can include extreme anxiety and panic.

One component of the public hysteria surrounding electricity erupted from the debate surrounding electrocution by virtue of alternating current (AC) vs. direct current (DC). In June of 1888, the then New York governor signed into law the Electrical Execution Act abolishing hanging as a punishment for heinous crimes and substituting death by electrocution. This despite the fact that the public was becoming increasingly horrified by the execution of dozens of dogs, calves, horses and cows by electrocution.

Even currently, more than 30,000 non-fatal shock accidents occur each year. Each year, there is an estimated 60 electrocutions associated with consumer products (ESFI 2014)

The Executioner's Current

An inventor, Alfred Southwick, adopted a strong personal belief that AC was superior to DC as a means for effective electrocution of human criminals. During the debate regarding New York State's new Electrical Execution Law which preceded the first execution of a human by electricity, Southwick was quoted as saying "The most effective of these are known as 'alternating machines,' manufactured principally in [the USA] by George Westinghouse. The passage of the current from these machines through the human body, even by the slightest contacts, produces instantaneous death." (Edison Archives 1887)

On August 6, 1890, William Kemmler was the first person to be put to death intentionally by electrocution. (Moran 2002) Westinghouse alternating current (AC) dynamos were used in the execution. This led to a public relations campaign which was part of an overall plan by Edison to discredit AC in favor of his DC system. Westinghouse attorneys sued prior to the execution in an attempt to prevent the use of the Westinghouse dynamos. (U.S. Patent Office) The attorneys filed a writ [a legal document filed with the court] for the return of the dynamos. The suit failed.

According to press accounts, the execution was a horrific affair. As it was the first intentional execution of a human being, the electrodes were not optimally applied and after an agreed upon 15-second jolt, Kemmler did not die—the dynamos had to be restarted and a prolonged 2000-volt charge was applied searing his flesh and creating a putrid odor.

Almost immediately, the medical men and the press began to quarrel over whom was responsible for the botched execution. (Moran 2002) For example, the *New York World Magazine* denounced the execution as "very cruel and very shocking." From all visible indications, according to the paper, Kemmler died in slow, torturous agony. The public were outraged. The *New York Evening Post* published an editorial responding to newly released pictures of the electric chair used. The editors denounced the execution as a medieval torture chamber.

If modern electric systems never evolved, then Edison himself would have contributed substantially to the demise of electricity. First he tried to hinder AC's progress through legal challenges, and then finally he appealed to the public through the media. Edison began to speak out about the "dangers" that he felt were inherent in AC. In February 1888, Edison published a pamphlet entitled "A Warning from the Edison Electric Company" arguing that AC was bad.

Death by Wire

In 1884, Edison declared in a speech that "unless the electric light companies reorganize their systems, so as to reduce the tension*, there is going to be trouble. The tension will become dangerous. It will kill men, and something like panic will follow." (*NY Evening Sun* 1889) While designed to discredit AC systems, it could be argued retrospectively that this likely discredited all electrical systems.

Another inventor, Harold Brown, believed that "New Yorkers needed to recognize that alternating current was simply too dangerous for residential or commercial use." He believed that even low-voltage AC must always be dangerous since "impulses are given first in one direction, then in the other several thousand times a minute." (Brown 1886)

*Edison was referring to the voltage levels which AC systems use to leverage the ability to transform among voltages and transmit longer distances.

In one of the first near disastrous examples of wiring a house—electrifying the home of J.P. Morgan in the fall of 1883 was also a near disaster. Soon after the system was installed, workers realized that "the house was pervaded by a strong smell of wet, burned wood and burned charcoal." (Jones, 2004) A short in the electric system caused portions of the internal structure of the house to smolder—but fell just short of destroying the home. This event suggested that the public's worst fears were confirmed: electricity was dangerous; caused fires; and could destroy homes and property.

There is further evidence that electricity's development could have been arrested even before it hardly started. An early example of the technological pessimism of this period was the electric wire panic that took place in New York City in the 1880s. At this time, the streets of the city were darkened from the skies by overhead electric wires. The fact that these wires were overhead, and not underground, along with debate over electrical safety, generated great fear of electricity. This fear reached its height in 1889 when a series of horrible electrical accidents terrorized the city. (IEEE 1995)

Then in 1888, the infamous "blizzard of 1888" howled through New York City with 60-mile-per-hour winds devastating buildings and destroying the electric distribution grids which sagged and collapsed under the ice. It was just one month later that the great public fear of electrical "death by wire" began (Jones 2004). In April 1888, a youth was electrified by one of the wires originally buried by the prior storm while walking down the street. The public was appalled. From then on, the term "death by wire" was popularized by the New York press. Only one month later, an electrical worker was killed, and citizens increasingly associated electricity with danger and death.

Howard Crosby, a so-called reformer (a term used for political activists at the time), declared that the wires were "A fearful source of death, and a constant menace to the lives of our fellow citizens. It would be far better for us to go back to the gaslights than thus to risk precious lives. The companies who make their fortunes by the

Figure 2-2. Overhead wires in New York City (IEEE)

electric lights seem to have no regard for aught but their purses." (*New York Tribune* 1889)

Some believed that new electric technologies would contribute to the moral and physical degradation of society. By the mid-1880s, there were 10 electric distribution companies in New York City with all 10 running overhead wires over the same street in some cases. Between May 1887 and September 1889, seventeen New Yorkers were electrocuted by overhead wire configurations.

(*New York Times* 1889)

"As science pushed aggressively to uncover nature's secrets, the public encountered, in magazine essays, newspaper articles, and fiction, disturbing characteristics of scientists as heartless, godless transgressors. And as electricity as a fearful phenomenon." (Simon 2004)

Wabash, Indiana

An example of the public's paranoia with electricity is the reaction to one of the first large outdoor electric lighting installations in the world. The city of Wabash, Indiana, USA, energized their installation of electric arc lamps.

"The people stood almost breathless, overwhelmed with awe, as if in the presence of the supernatural... The strange, weird light, exceeded in power only by the sun, rendered the square as light as midday. Men fell on their knees, groans were uttered at the sight and many were dumb with amazement. It drove the darkness back and out of the entire city of Wabash so that now the people could see to read on nearly all of the city's streets by night." (Ksander 2008)

Wabash was the first city to be lit solely by electricity and to own its own municipal power plant (a small dynamo driven by a threshing machine engine). In 1876, Charles F. Brush invented a new type of simple, self-regulating arc lamp, as well as a new dynamo designed to power it. Earlier attempts at self-regulation had often depended on complex clockwork mechanisms. The simpler design by Brush Electric Light Company used a solenoid combined with a clutch mechanism to adjust the carbons over their entire length. In the late 1870s, Brush arc lamp installations were being purchased by individuals, department stores, theaters, and factories.

A local newspaper, *The Wabash Plain Dealer* reported the concerns expressed by various citizens expressing about the use of artificial illumination. Concerns were that chickens only sleep when it is dark, and if the nighttime were illuminated, chickens would die. In addition, the newspaper claimed that illumination at night would enable nighttime raccoon (coon) hunting. Concluding that this would drive up the price of coon hounds (dogs which are used to hunt coons). (Friday Night Lights)

The War of Currents

In the late 1880s, a communications battle referred to as the "War of Currents" (also referred to as the Battle of Currents) raged between George Westinghouse and Thomas Edison based on a clash caused by Edison's promotion of direct current (DC) and Westinghouse's promotion of alternating current (AC) power systems. The War of Currents heightened the public's concerns over electricity. And could have contributed to the demise of the modern power system. While merely a communications battle, it has been observed by many historians that the war of electric currents was one of the most vicious battles in American corporate history. (Jones 2004)

During the initial years of electricity distribution, electrification and the expansion of systems, Edison's direct current was popular in the U.S. Edison's DC was literally the first electric utility and the first functional power system in the world. DC easily powered incandescent lamps and electric motors. Direct-current systems could be directly used with storage batteries, providing valuable load-leveling and backup power during interruptions of generators. Direct-current generators could be easily paralleled, allowing economical operation by using smaller machines during periods of light load. At the introduction of Edison's system, practical AC motors were not yet available. In addition, Edison had invented a DC meter which allowed customers to be billed for energy consumption, but this meter worked only with direct current—further disadvantaging AC at the time. AC systems would prove to be superior through the ability to use transformers to step up and down voltage. The only practical transformer technology at the time was a relatively primitive open-core bi-polar transformer. The transformation efficiency of the early transformers were low and contributed in part to the slow adoption of AC systems. Early AC systems used series-connected power distribution systems, with the inherent flaw that turning off a single lamp (or the disconnection of other electric devices) affected the voltage supplied to all others. For many readers, this may be reminiscent of older Christmas tree lights. (Wikipedia 2013)

Eventually, Edison unleashed the full force of his monumental fame and prestige to attempt to persuade the public that electricity was safe, especially when buried, as was his DC system. Edison contrasted the DC system with the dangers of Westinghouse's AC system which was strung on overhead wires. Ultimately, the flames of the so-called "Current Wars" were fueled by Edison's former employee and AC inventor and proponent, Nicola Tesla. Tesla declared that Edison's technology was passé, obsolete and primitive. (Jones 2004)

The World's Columbian Exposition held in Chicago in 1893 proved another obstacle in creating and maintaining a positive impression of electricity. In the bids for the fairgrounds, General Electric bid $18.50 per lamp and the only competitor bid was $6.80 per lamp. The competitor turned out to be a representative for Westinghouse. (Jones 2004)

Safety Concerns

Interestingly, consumer concerns about safety increased which could have inhibited electricity's adoption. Some of these linger still today. The Electric Power Research Institute (EPRI) undertook considerable market research to develop a system called CLASSIFY. (EPRI 104567) EPRI's CLASSIFY System was actually designed to give utilities the tools they need to understand the needs and attitudes that drive their customers, so that they can be more responsive to them. EPRI research identified a total of 11 different needs that appear to be most responsible for driving consumer energy-related preferences.

The 11 basic needs that drive residential energy decision-making are outlined below. Note that individual customer's ranking differed on each of the 11 basic needs, depending on his or her preferences and behavior regarding energy-related issues.

Note that three of these are directly related to safety concerns: surge protection, enhanced security, and safe appliances.

- *Low Energy Bills*—Customers with this need are concerned about controlling their energy operating costs. It is very important to them to look for ways to reduce their energy bills, including, for example, the substitution of warmer clothing for higher thermostat settings in cold weather.

- *Increased Comfort*—Some customers indicated a strong need to maintain a comfortable home through the use of heating, cooling and dehumidification appliances. In selecting their heating/cooling systems, they were very concerned about the evenness of the temperature throughout their entire home.

- *Surge Protection*—This need reflects some customers' concerns with power surges. They are particularly afraid that a sudden increase in electrical voltage could damage their appliances and have a need to protect themselves from the problems that would result.

- *Time-Saving Appliances*—The convenience need reflects an interest in using time-saving appliances, such as dishwashers and microwaves, to make more time available to spend on other activities. Some people also like the fact that these appliances reduce the labor involved in cooking and housework. Simplicity of use is also seen as adding to convenience.

- *Resource Conservation*—The driving force behind this need is to decrease the environmental impact of electricity usage and protect our natural resources. Customers exhibiting this need are concerned about equity issues, such as using only their fair share of electricity and preserving resources for future generations. They believe that they have a social responsibility to reduce their electrical usage.

- *Enhanced Security*—Customers with this need are concerned about their own personal security and that of their home and property. In an effort to discourage theft and vandalism, they keep lights on in their homes even when they are away. Outdoor lighting increases their sense of security and personal safety.

- *Safe Appliances*—A high score on this need reflects concerns about the safety of electric and gas energy, both inside and outside the home. Consequently, some customers are concerned about the possibility of downed power lines, gas leaks, and the safety of their electric and gas appliances. They worry that their appliances could shock users or cause a fire. Their safety issues include fear of electrical discharge from microwave oven and TV sets.

- *Personal Control*—Customers with this need like to control their own appliances, using them to accommodate their needs for comfort and convenience. Such people are uncomfortable with the thought of losing control and would, therefore, be unlikely to volunteer to have a utility restrict their energy use.

- *Attractive Appliances*—When purchasing an appliance, customers who feel this need strongly care more about its aesthetics than its performance. Such a customer would probably worry more about whether an appliance enhanced the décor of a home than how it was rated by *Consumer Reports*.

- *Hassle-Free Purchases*—Customers who feel this need strongly do not want to devote much time to researching or shopping for appliances, nor are they interested in specials or rebates. At the other end of the spectrum are customers who enjoy shopping for the best values. They are willing to devote the time to researching features, reliability records, conservation options, and purchase price, in order to acquire an appliance that can meet their requirements.

- *High-Tech Appliances*—Customers with this need are interested in having the latest technology, with lots of features and options. Gadgetry is often preferred over ease of use. They are attracted to more of the features offered with VCRs, computers, and tape decks than the average consumer.

Not all of the 11 needs are equally important to all people. While individuals obviously vary in the importance they place on each need when making decisions about appliance purchases or energy usage, there are also clear patterns that emerge in needs importance when you look across all residential customers.

The information plotted in the chart shown in Figure 2-3 represents the relative importance that residential customers as a whole attach to each of the basic 11 needs. The needs are listed in order of their average importance to the population at large. A need with a score plotted farther to the right on the scale is more important, while one with a score in the middle can be viewed as average in importance, and one with a score to the left of center can be thought of as less important. Note that surge protection and security rank among the highest.

Electricity in buildings and power systems still isn't 100% safe. In the USA there are 47,700 electrical home fires annually, 2,400 children are injured by tampering with electrical receptacles, and 2,000 people are injured in personal property electrical fires—some directly attributable to electricity. Most tragically, there are 480 deaths annually in personal property electrical fires and numerous deaths from carbon monoxide poisoning—some related to electrical fines and others which could have been prevented by the US of CO detectors. (NEMA May, 2014)

The Failure of Electricity as We Know It

Mounting consumer paranoia about electricity and its dangers, its high price and unreliable service could have created

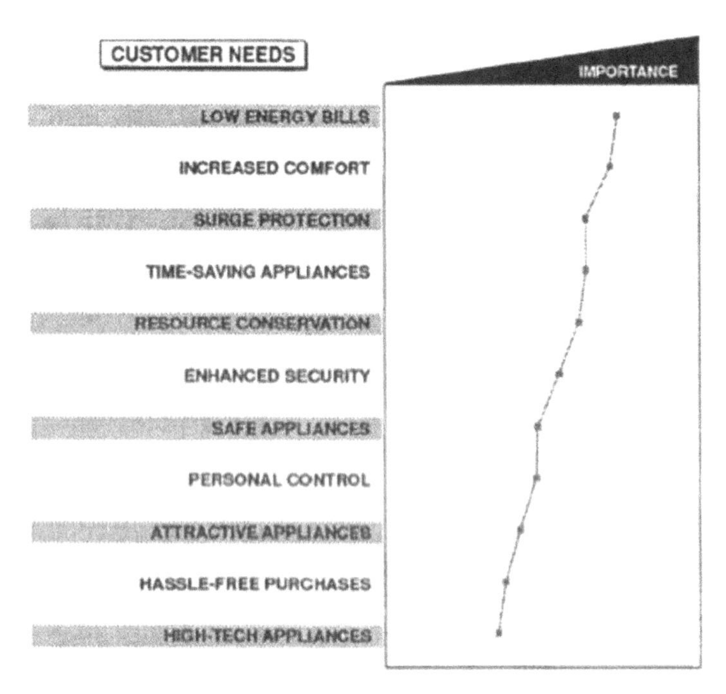

Figure 2-3. Residential Customers Overall (EPRI 104567)

an insurmountable marketing challenge for electricity companies. What if, over the months ahead, following its commercial introduction, new customer signups diminished and existing customers dropped like flies? As investors witnessed the virtual market failure of the retail electricity business, the resources needed to move the electricity business forward dried up. One can imagine that it then never regained traction.

Post Script: Good or Bad?

Technology is not intrinsically good nor intrinsically bad. Technology based on electricity is instrumentally good to the extent that technological devices are used to facilitate the creation of value. Technologies that help humans save energy or reduce waste are inherently good. Other technologies which are beneficial to humans are also good: these include those that help humans live longer, healthier lives; those that provide shelter; or those that facilitate interpersonal communication.

Chapter 3

Living Without Electricity

One way to estimate the value of electricity is to consider living without it in today's modern world. Anita Evangelista published an article related to this point in a 2002 edition of *Backwoods Home Magazine*. Her article focused on the issue of surviving a long-term blackout. She posited that there are five primary areas of concern if the power goes off: light, water, cooking, heating/cooling and communication. Using Ms. Evangelista's outline, supplemented by other references, the author summarizes how someone would temporarily live without electricity in the modern world.

1. *Lighting*: The simplest form of emergency lighting is a flashlight. Well-prepared homes have one in virtually every room as well as at least one battery powered lantern in a central location. Rechargeable batteries can be very useful in a prolonged outage, provided solar-powered or car chargers are available—albeit they are limited in light output. Candles may also be used, but it would be necessary to stock up ahead of time. It also takes several candles to provide sufficient illumination to do many tasks such as reading. The oil lamp—one which burns kerosene—is preferable over candles, and provided sufficient kerosene is stockpiled, can be superior to battery-powered devices. In the case of candles or oil lamps, it will be necessary to also have matches on hand. And, any open flame is dangerous!

2. *Water*: Water supply subsequent to a blackout may not be immediately interrupted. But, eventually, in most municipal water systems, the tap will run dry. Obviously, the easiest

way to guard against water shortages is to store water. Several agencies suggest one gallon per person, per day. Accessing streams or drilling wells and extracting water with hand pumps are also options. Purification filters can allow access to marginal water supplies or swimming pool water.

3. *Cooking*: Heating food is not essential unless it is in regard to uncooked meat or fish. Outdoor cooking is always an option and the first priority in preparing meals after losing power, under circumstances where the outage is expected to be prolonged, is to cook the food in the refrigerators and freezers which will spoil. Obviously, homes equipped with propane-fired barbeques can serve households well under these circumstances. Kerosene- or natural gas-fired refrigerators are also available and can be very useful, albeit expensive to have on hand in the event of a prolonged blackout.

4. *Heating and Cooling (Space Conditioning)*: Heating and cooling—predominantly heating—can become a priority in the event of prolonged absence of electricity. Clearly heating, as winter temperatures dip, is the most serious need. Numerous alternatives are available including wood stoves, propane or kerosene heaters, or natural gas heaters without electric ignition.

5. *Communications*: The issue of electronic communications would only come forward when living without electricity in the context of a blackout is discussed. If the electricity enterprise never evolved, then electronic communications would not have evolved. However, once society experienced communications, then its absence would be traumatic. Keeping connectivity in a blackout depends enormously on the extent and duration of the blackout. Homes with photovoltaic power systems which have electric storage and proper controls can survive nicely through the blackouts. During a black-

out, and in the short-term, cell phones can be charged with car-powered devices or solar chargers—until the cell towers are out of service. Homes can be supported with plug-in electric vehicles or hybrids, until the supplemental natural gas or gasoline-driven generators can no longer operate.

During blackouts, the prospects for getting power restored seem more remote, there are other options which may suddenly become more reasonable than they had been. Examples include canning, use of solar ovens, or making a fabricated toaster from large coffee cans.

According to some, living without electricity is actually a lot easier than most people think. "How can I not have TV? Or the blender, mixer, lights? Well, many people still choose to live without all the hassle, electric bills, etc. And life actually becomes easier, slower, and even more serene. You go to bed earlier, so you get up earlier in the morning. You are more in tune to the sun, the seasons and your life. Do you know what the weather is today? Look outside! What time is it? Time to get up because it is dawn. Do you need to stay up until midnight to watch TV? Nope—just to sit outside around a fire and watch the stars…" (High Lonesome Ranch 1995)

HOW IS ELECTRICITY USED TODAY?

Another perspective on the hypothesis of living without electricity can be gained by reviewing each sector's use of energy in the period just prior to the potential establishment of functional electric power systems which made electricity virtually ubiquitous in the western world. There are considerable differences between urban households and households in suburban and rural areas.

How would society have replaced the uses elucidated in Figure 3-1?

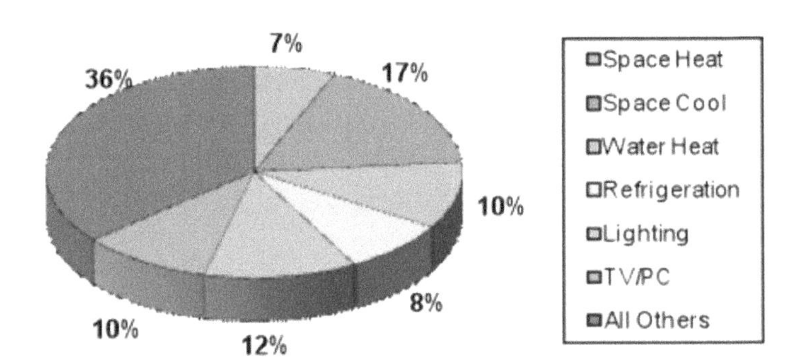

Figure 3-1(a) Residential Electricity Use by End Use—2013 (DOE/EIA Annual Energy Outlook 2013, with Projection to 2040, Tables 4 and 5 of Reference Case, model run ref2013.d102312a)

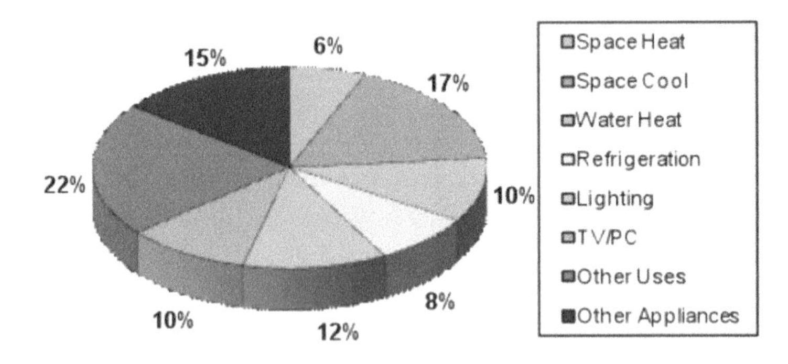

Figure 3-1(b) Residential Electricity Use by End Use—2013—with Other Appliances (Ibid)

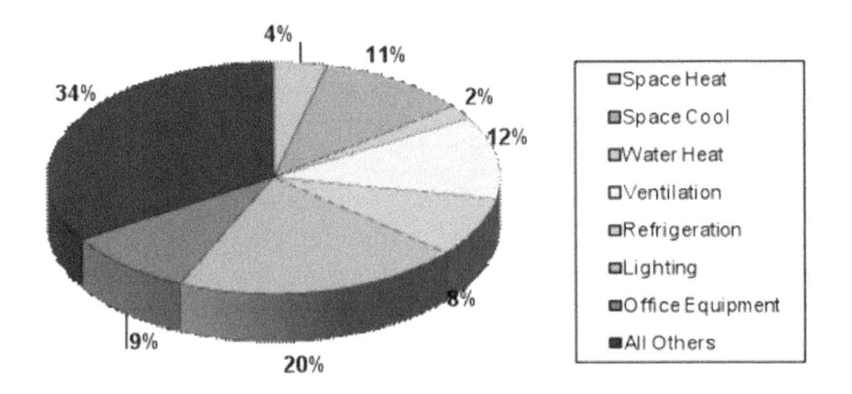

Figure 3-1(c) Commercial Electricity Use by End Use—2013 (Ibid)

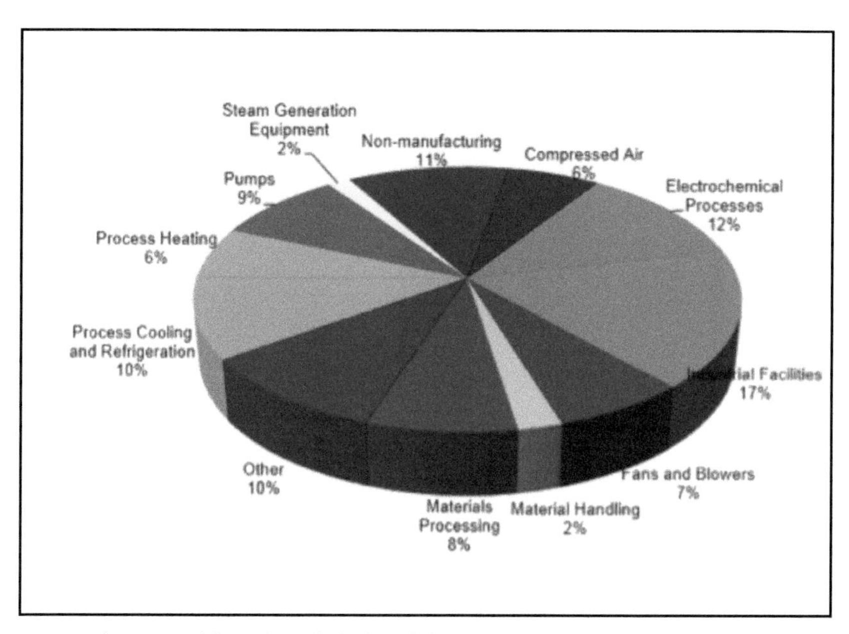

Figure 3-1(d) Industrial Electricity use by Process—2009 (Ibid)

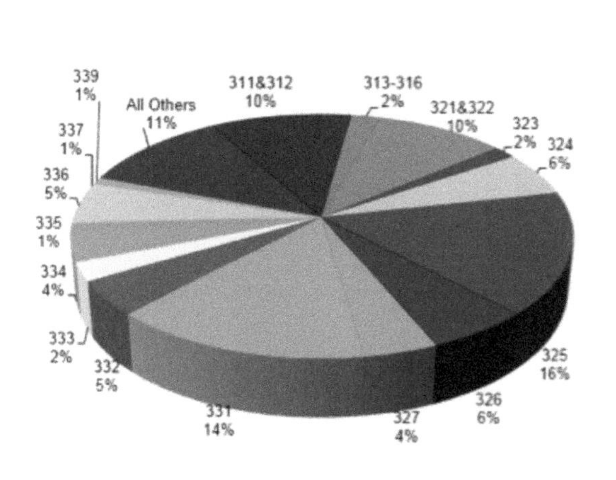

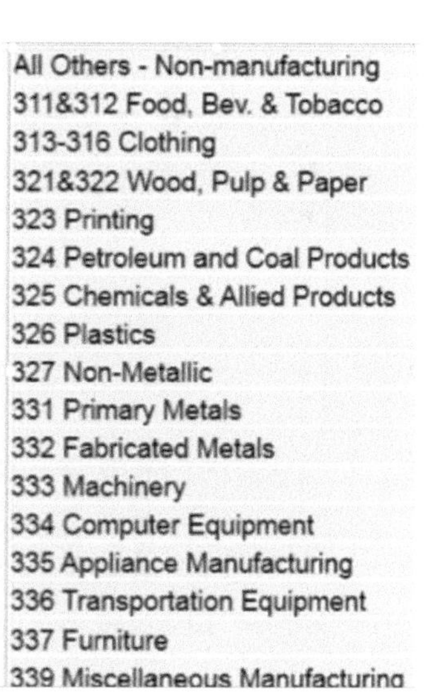

Figure 3-1(e) Industrial Electricity use by NAICS code—2009

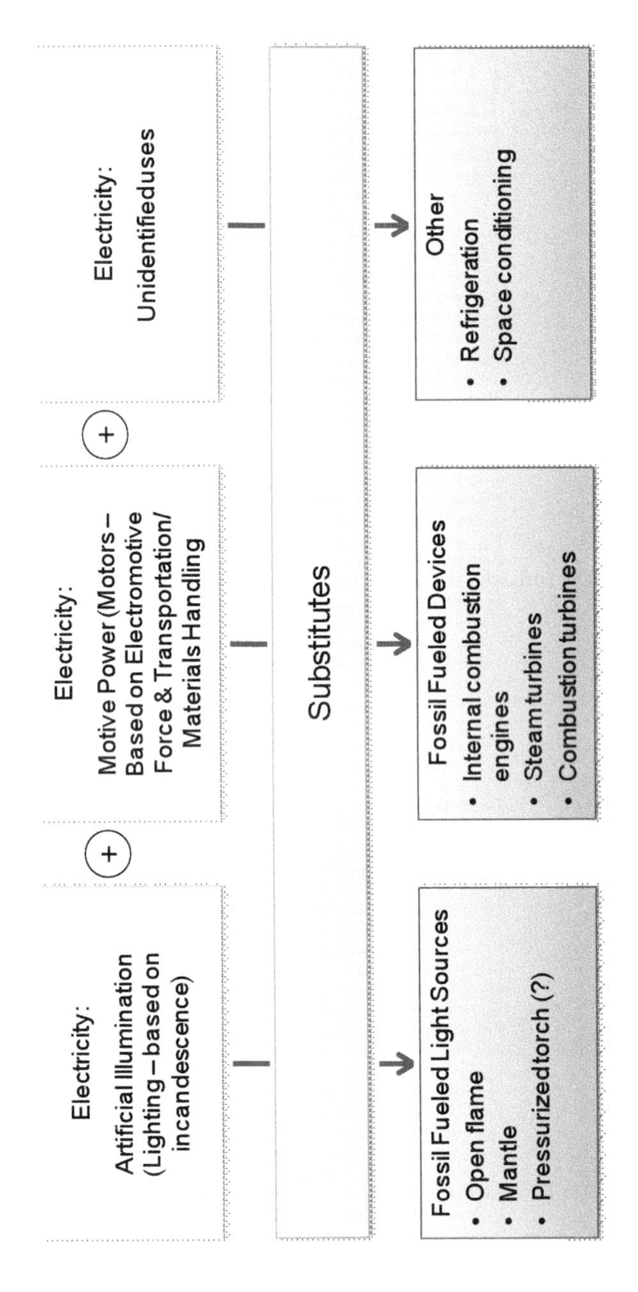

Figure 3-2. End Uses of Energy in the 1800s: Electricity and Substitutes

Electricity in the Domestic Setting

The most obvious example of electricity's value is to simply review how it has changed life in the home. Today, electricity powers homes, enables communications and household entertainment.

As the electric industry evolved, so did the technologies which could be found in each room of the house. The electric kitchen increasingly featured stoves, hot plates, broilers and water heaters. The laundry adopted electric washing machines and irons. Carpet sweepers could be found in every parlor.

The invention of electricity irreversibly altered family social and economic structure. Prior to the introduction of electricity, both men and women performed "domestic work." That is, because a majority of the families lived an agrarian lifestyle, daily tasks centered on the home and farm. Every member of a family performed specific chores. Men and women held equal levels of responsibility and both were valued equally for the work they did—their contributions. Most tasks—whether household or farm chores—demanded a great deal of physical energy and time from both men and women.

Prior to the late 1880s, everyone in the family "worked," and the literature suggests that they worked very hard! Each member contributed to the welfare of the family. Everyone, from an early age, worked for money, for credit in barter, or for room and board. Each type of labor performed by both sexes was considered work, even if currency did not change hands. Many tasks were "pooled" by a community. For example, a community often maintained a gristmill that all could use to process grain or it would band together to accomplish tasks such as constructing houses and barns, or harvesting crops. The community understood the absolute necessity of working together to survive and prosper. (Gellings 1994)

The Human-powered Home

There is a practical limit to the extent to which human power can displace electric-powered devices. This limit is based on how much power and energy a human is capable of. Power is the force that can be applied at a point in time; the intensity of which dictates the ability of the force to act. Energy is the amount of overall power that can be applied over time.

Work occurs when something is displaced by power. Energy equals power x distance measured by Newton-meters or joules and power is work/time or joules/second measured in watts. Since one calorie = 4,186 joules (USDA) and active adults consume 2,200 calories/day—a human can exert 9.2 x 106 joules/day. A body can only exert 1/3 of those joules to do work or about 100 to 300 watt-hours/day.

Value of Electricity: Human power equivalent
Hypothesis: Estimating the value of electricity by substituting human calories
Assumptions: Humans produce 200 kWh/day using 2200 calorie; eggs 155 calories @ 25 cents
Value of Electricity: $.0972/kWh

Human power can never approximate the power which electrically powered devices and appliances can produce.

Electricity has fundamentally changed the ability of humans to do work. Early man had the benefits of human muscle alone. Human muscle is equivalent to an electrical power of approximately 35 watts.

Before the industrial revolution, every tool was human-powered. "Rock and stick served as hammer and lever. Other simple machines—wedge, pulley, wheel, inclined plane and screw—followed." (Dean 2008)

There is ample evidence that humankind would have developed mechanical devices or robots which could have replicated some of the advanced electric devices. A review of historic developments demonstrates prospective innovation. For example, the Canard Digérateur or Digesting Duck was an automated duck invented in 1739 and reportedly had the ability to eat kernels of grain from a simulated mouth and produced simulated feces from

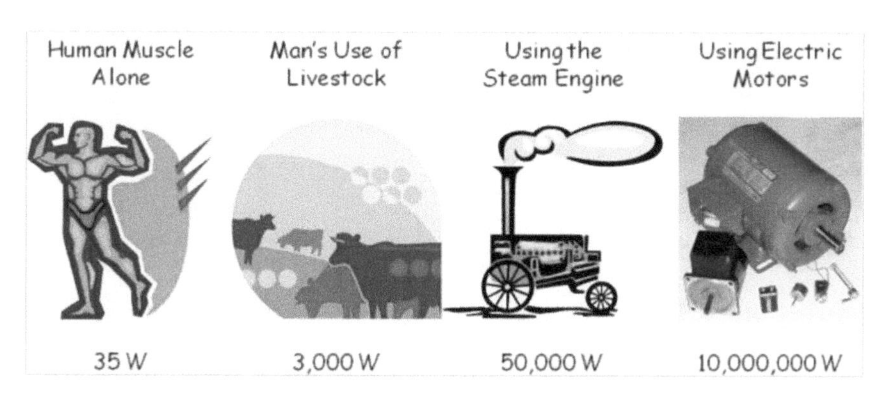

Figure 3-3.The Power of Human Muscle as Compared to Other Technology

a second orifice. A three-doll autonatron was built consisting of a phenomenally functional set of dolls which could even write text. A final example is the "Turk," an automated chess player which appeared to play a real chess game against a human opponent. Street traffic signals and similar mechanical needs could possibly have been accommodated by devices such as these without electricity.

Human ingenuity has always been present in devising technologies which would harness human energy. Early examples include the citizens of Prague building a human-powered crane. Other inventions soon followed such as a horse-powered seed drill and the spinning jenny which was used to spin yarn. Eli Whitney's human—powered cotton gin soon followed based on similar concepts.

The first recorded human-powered device was a potter's wheel. This is not a surprise, but it reflects the three elements of human-powered devices—treadle, continuous belt and flywheel. (Dean 2008) Other inventions included more commonplace household machinery such as the treadle-powered sewing machine. The importance of the sewing machine has been cited for the hours it added to women's leisure. The Singer Company sold 810 such machines in 1853 and 500,000 in 1880. (Hounshell)

Eventually the concept of foot-powered devices extended to the bicycle and bicycle-powered devices. These included many

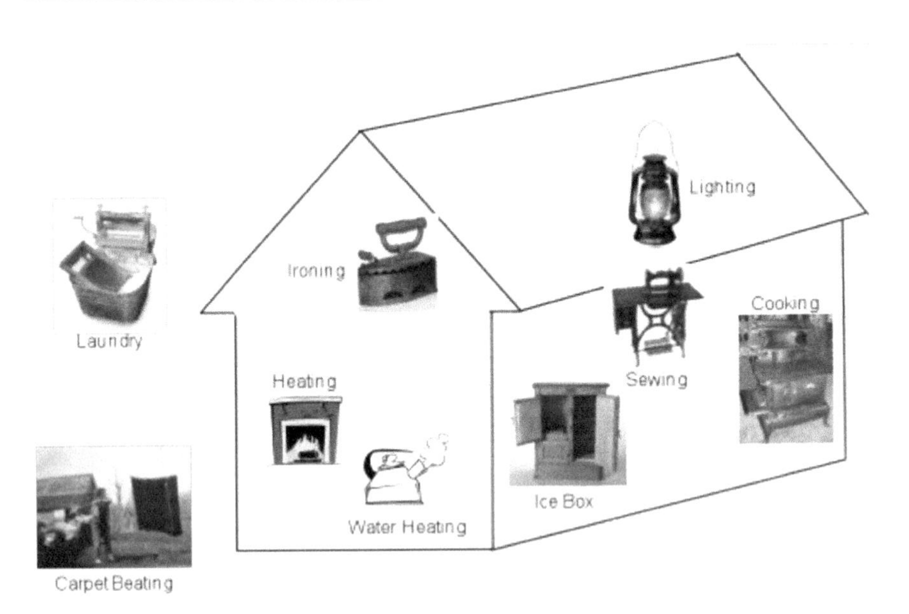

Figure 3-4. Non-Electricity Household Energy Use in the 1880s

devices which leveraged human legs as an energy source. Among these inventions, Table 3-1 lists a number of human-powered devices.

It has been offered that human-powered technology is more "appropriate" than that powered by other energy sources. It is rumored that even Gandhi didn't oppose everything modern, but insisted on choosing technology that would help the poor work themselves out of poverty. Devices that helped the poor worked themselves out of poverty was based on local human power.

Leveraging Electricity to Aid Humankind

Humankind's ability to harness energy followed a timeline of over 10,000 years. (See Figure 3-5)

Humans discovered fire and began to harness its power for warmth and cooking at about 10,000 B.C. It wasn't until 1776 that man learned to really control it by virtue of the development of the safety match. Only one later, in 1777, fire was used in furnaces

Table 3-1a. Human-powered Devices of the Late 19th Century (Dean 2008)

Hand-Powered	Hand-Powered	Hand-Powered	Foot-Powered
Apple parer	Baler (or bailing press)	Bean sorter	Blocking machine (for hat making
Blower (for blacksmithing)	Bobbin winder	Bone cutter	Boring machine
Book trimmer	Boring machine	Box nailer	Broom winder (broom-making machine)
Butter churn	Cement mixer	Centrifuge	Cigar maker (or former)
Chaff cutter	Cheese press	Cherry pitter (or stoner)	Circular saw
Cider press	Coffee mill	Copy machine	Dentist drill
Corn sheller	Cream separators	Crimping iron (for pleating skirts)	Drag saw
Cultivator (push-type)	Dishwasher	Drill press	Drill press
Edging machine (for hat making)	Eggbeater	Eggbeater drill	Former (shaper)
Fanning mill	Food chopper	Food mill	Graphophone (dictating machine)
Forage cutter	Fruit press	Glass-cutting lathe	Harvester (treadmill-type)
Grain huller	Grain mill	Grain roller	Hook and eye maker
Grape crusher	Hat printer	Ice cream maker	Knitting machine
Ice crusher	Jeweler's lath	Jeweler's polisher	Milking machine
Juicer	Knitting machine	Laboratory shaker	Miter saw
Lard press	Lawn mower (push-type)	Marmalade cutter	Mortising machine
Meat grinder	Nut roaster	Oil press	Printing press
Oilseed cake crusher	Olive pitter	Package bundler	Punch machine

Table 3-1b. Human-Powered Devices of the Late 19th Century (Dean 2008)

Hand-Powered	Hand-Powered	Hand-Powered	Foot-Powered
Paint grinder	Paper cutter (used in book binding)	Pea sheller	Rip saw (table saw)
Pencil sharpener	Phonograph	Picket maker	Riveting machine
Planer (hand-cranked)	Post drill	Poultry delouser	Screw-cutting lathe
Printing press	Radio transceiver	Raisin seeder	Scroll saw (jigsaw)
Record Player	Rip saw (table saw)	Root grinder (or turnip pulper)	Seed cleaner
Root washer (or potato washer)	Rope maker (rope weaving machine)	Rotary hoe (push-type)	Sewing machine
Rotary sickle (push-type)	Rotary tiller	Sausage press (sausage stuffer)	Spinning wheel
Seed broadcaster	Seed cleaner	Seed potato cutter	Stave cutting machine
Seed spreader	Sewing machine	Stamp mill (for beating rags to make paper)	Thresher
Steak crusher	Sugar press	Tenoning machine	Tip stretcher (for hat making)
Ticket printer (ticketing machine)	Tire shrinker	Tabaco shredder	Tool sharpener
Tool sharpener	Vacuum cleaner	Washing machine	Typewriter (with treadle return)
Water pump	Well driller	Whale blubber mincing machine	Vegetable bundler (or buncher, for asparagus or celery)
Wringer (mangle)			Weaving loom
			Wood lathe

to commercially produce iron. Fire again served man by the invention of the steam engine in 1826.

As man progressed he began to harness the use of livestock. The typical animal, under a man's tutelage had power equivalent to 3,000 watts. As the industrial revolution matured, and the steam engine appeared, with it one man could harness approxi-

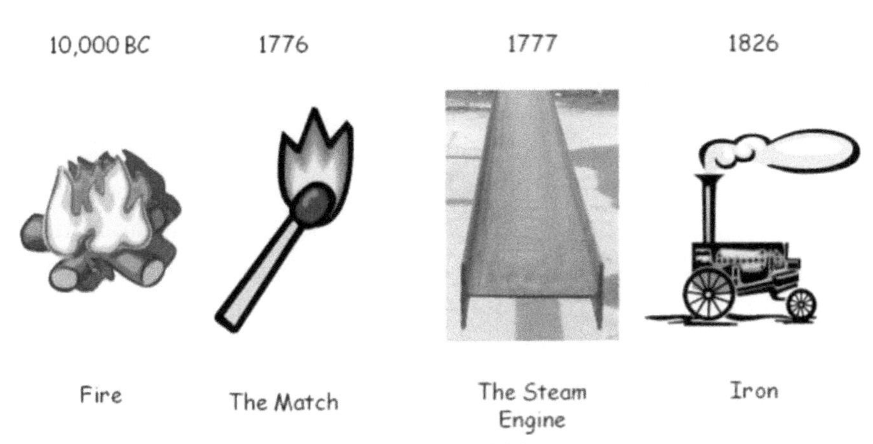

Figure 3-5. Timeline of Human Ability to Harness Energy

mately 50,000 watts. But it wasn't until electricity that man could harness nearly unlimited power. For example, a large motor of 10,000 horse power could allow man to harness the power equivalent to 10,000,000 horses or more.

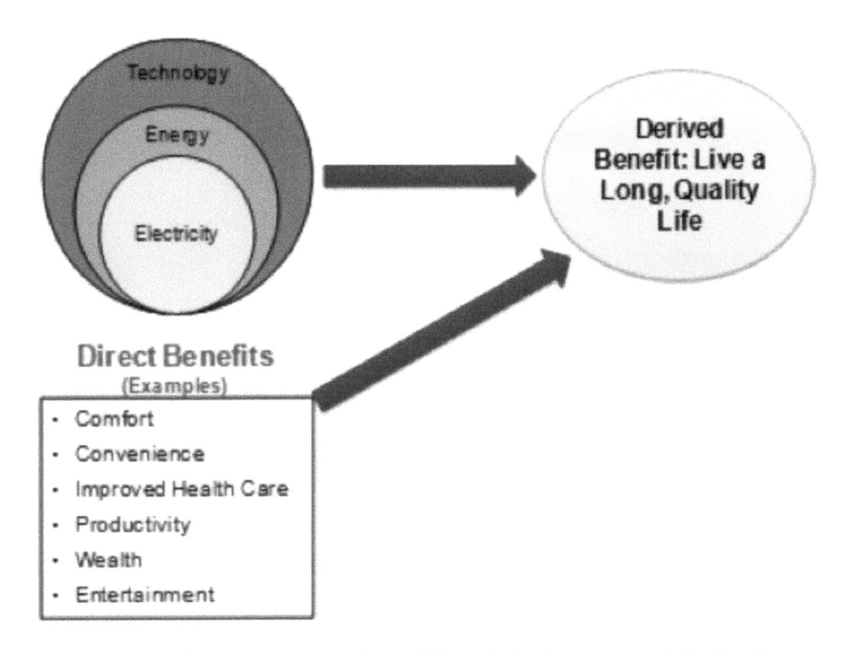

Figure 3-6. Illustrates the Value of Electricity, Energy and Technology

Commercially Distributed Electricity

Electricity was first commercially distributed during the period 1879 to 1881. In the beginning, towns used it only for artificial illumination, especially streetlamps. As electric technologies evolved, inventors and utilities began understanding its potential applications in the domestic "workplace" as well as in industries. By 1890, small electric motors were available that could power household appliances. This paved the way for numerous applications. The electric iron was the second major residential appliance (after lighting) to achieve widespread use. In 1908, the first vacuum cleaner was introduced by Hoover.* Washing machines and refrigerators were available soon afterwards. Although, refrigerators were not effectively mass marketed until the 1930s.

By the 1920s, nearly two-thirds of U.S. households used electricity to power lights and appliances. At first, only wealthy and upper middle class households could afford new electric ap-

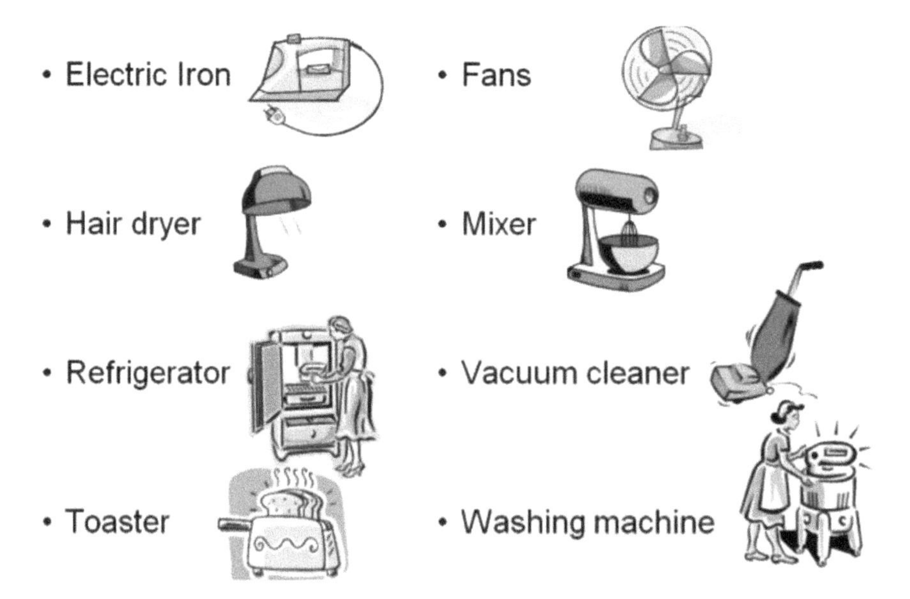

Figure 3-7. Popular Household Electrical Appliances Invented in the 1890s to 1920s

*The function of vacuuming is still referred to as Hoovering in the U.K.

pliances. As costs of production and distribution declined, new appliances became affordable to both middle class and poorer households.

As shown in Table 3-2 even in 1940 many electric technologies were still not used widely in U.S. homes. By 1960, nearly 100% of households in the U.S. had electricity. Table 3-3 details the major electric appliance used in 1987. These appliances displaced or substantially altered many traditional household functions and activities. Table 3-4 depicts use as documented in 2012.

Table 3-2. Illustrates some of the household uses in the 1940s

Household Uses	Percent	
	1940	1960
Bathtub or shower	61	88
Wood, coal or oil stoves	45	5
Wood or coal for heating	78	16
Electricity	79	99

As electric technologies proliferated, electricity users began to recognize electricity's ability to provide a clean energy supply and contribute to cleaner living conditions. Electricity use eliminated the dirt and grime associated with burning coal and wood. It also facilitated food preparation, storage, and cleanup. Medical researchers increasingly recognized the link between disease and unsanitary conditions as well as the ways electricity use could solve these problems. Americans linked cleanliness with moral decency and good behavior. Thus, as washing garments became easier with machines scrubbing them in hot water using strong detergents, people changed clothes more frequently.

Prior to the adoption of electric lighting, homes using gas or kerosene for illumination would typically need to repaint their ceilings each year in order to mask the soot which was produced by burning fossil fuels.

Table 3-3. The Evolution of U.S. Household Electric Appliance Use (Gellings 1994)

Type of Appliances Used	Percent of Households Using Electric Appliances by 1987
Television Set (color)	92.7
Television Set (b/w)	35.8
Clothes Washer (automatic)	73.3
Range	56.8
Oven	56.6
Clothes Dryer	50.7
Dishwasher	43.1
Window or Ceiling Fan	46.2
Microwave Oven	60.8
Water Heater (for one household's use only)	33.7
Air Conditioners (room)	30.8
Electric Blanket	30.0
Air Conditioner (central – for one household's use only)	32.5
Freezer (not frost-free)	23.0
Humidifier	14.6
Freezer (frost-free)	11.7
Waterbed Heater	13.9
Dehumidifier	10.0
Whole-House Cooling Fan	9.5
Swimming Pool/Jacuzzi/Hot Tub Heater	0.7
Refrigerator – 1	86.2
Refrigerators – 2 or more	13.6

Table 3-4. Current U.S. Residential Appliance Saturation (U.S. Dept. of Energy, Energy Information Administration, AEO 2012

Percent Households	Type of Appliance
99%	household with televisions
99%	households with refrigerators
82%	households with a clothes washer
79%	households with a clothes dryer
79%	households with DVD players
76%	households with at least one computer
61%	households with central air conditioning
59%	households with a dishwasher
50%	households with a DVR

There are several energy-related functions that all modern households undertake. These include lighting; meal preparation; cleaning; space heating; laundry; cold food storage (short-term food preservation); and sewing. Urban and rural households may also have long-term food preparation (e.g., canning and cold storage); tending to animals (e.g., chickens and livestock); gardening; provisioning heating and cooking fuels (e.g., wood, kerosene, etc.); and certain related uses like blacksmithing, cabinet making, and cooling milk in dairy farms.

The Value of Time—Labor Saving in the Home

Many believed that electrification of the home would save labor from housework. Electrification of rural areas provides the best data to assess this issue. The data shown here was collected during the 1920s as studies of farm life became popular. At that point in time, many rural areas were just being electrified. At first glance, ownership of new household technologies did not appear to reduce the time spent on household tasks. (Kline 2000)

However, more careful analysis showed that these devices seldom reduce the amount of work, but they "merely enable people to accomplish more with the same effort." For example, the sewing machine just allowed women to sew more. And other studies showed that electric appliances saved time women spent on household chores—however, they simply used that time to do more on the farm. (Kline 2000)

Table 3-5. Sample Daily Time Record of Oregon Farmhouse (Approximate) (Based on Kline 2000)

Task	Minutes Spent
Grooming	80
Meal Preparation	185
Serving Meals	80
Washing Dishes	40
Tending to Milking	55
House Cleaning	105
Child Care	105
Sewing	70
Tend to Fire	20
Tend to Garden/Farm	60
Sleep	455
Clothes Washing	60
Leisure (read, radio, phone, converse)	105

Examining the technologies that were in place in the late 1880s allow a glimpse at what life would be like without electricity. Of course, the non-electric technologies which provided ener-

Value of Electricity: Increasing household ability to increase productivity or leisure time
Hypothesis: Electrification enabled homemakers to do more or to increase leisure time
Assumptions: Electricity saves 225 Mins/day @ $7.00/hour/Avg. human 35 W/115.2 Million households
Value of Electricity: $3.34/kWh

gy convenience in 1880 have, indeed, evolved. This can be seen in analyzing households and communities which have chosen not to buy or generate electricity or to even connect to, the grid. This is true of most Amish communities in the U.S. and in similar isolationist communities elsewhere in the world.

Amish Use of Electricity

The Amish do use energy and often electric energy but remain largely disconnected from the grid. But how do the Amish do it? One of the best experiments on what life would have been like if the electric infrastructure never developed is to study the Amish communities in the United States.

The Amish have not rejected electricity but have largely rejected connecting to the grid.

Amish homes stand visibly unique from homes in a typical urban community in that there are no visible power lines, visible telephone or cable TV lines or TV antennas. As electrification continued to embrace rural communities in the 1890s, the large majority of Amish leaders felt that not connecting to power lines was consistent with Amish beliefs. Most Amish communities have accepted some battery-powered devices such as clocks, watches, flashlights, calculators, buggy lights, electric livestock fences, and photovoltaic-powered agitators in milk tanks. (Scott 1999) A few communities have embraced cell phones. Some also use electric generators to power shop tools and to recharge batteries. And they do use landline telephone service from phone booths. Of course, the Amish generally shun the use of technology—whether or not it is electrically based. Table 3-6 summarizes the use of technology in various Amish communities. This includes various forms of energy applied as substitutes for electricity.

Table 3-6. Technology Use in Various Amish Communities—Percent of Farms/ Households (Scott 1999)

Technology	Percent
Pressurized Lamps	90
Propane Gas	30
Mechanical Refrigerator	40
Motorized Washing Machines	97
Flush Toilets	70
Running Water/Bath	70
Power Lawn Motor	25
Rototiller	20
Chain Saw	75
Tractor for Belt Power	70
Tractor for Field Work	6
Pickup Balers	50
Mechanical Milkers	35
Bulk Milk Tank	35
Pneumatic Tools	70

Lighting

Candles—Many Amish homes use candles—albeit usually those with a more dense wax made from paraffin. These candles produce less soot and smoke than cheap wax candles. Paraffin is actually a by-product of petroleum refining which is heavily dependent on electricity. So by using candles, people are still dependent on electricity, albeit not electricity at the point of end use.

Battery-Powered Lamps– The Amish depend heavily on battery-powered lamps. These range from simple flashlights to more elaborate fluorescent lamps often powered by D-cell batteries. Providing lighting in the home with batteries can become quite expensive since batteries used every night will seldom last more than one week. Obviously rechargeable batteries are an option; however, an electric source is needed to recharge them. Lamps which have built-in photovoltaic cells for recharging are usually called solar-powered.

Oil/Kerosene—Oil and kerosene lamps have been widely used by campers and those who live off the grid. Their configuration, efficiency, effectiveness and cost vary considerably and can give off soot and smoke, heat and can be dangerous. Modern pressurized camp-style lamps can give off a lot of light and burn all night on a quart of fuel. Again, these lamps do indirectly depend on electricity as their fuel is a product of oil refining.

Pressurized Gas Lamps—While some Amish still use the passive kerosene lamps that were developed in the mid-1800s, most are now using the considerably more efficient pressurized gas lamps. These lamps use naphtha or "white" gasoline which is forced into incandescent mantles by compressed air. So called "city gas" or what is now natural gas was available and used in lamps. Some others produced gas from gasoline and piped the fuel into their houses. For outdoor lighting, the Amish use the same lanterns as do campers—pressurized mantle lamps. They can produce the same illumination as a 100 watt electric incandescent bulb. (Scott 1999)

Gas Lights—Natural gas and propane can be used in the home for lighting in a relatively safe manner. In either case, plumbing can be used to "modernize" the system, and the resultant lighting is rather economical. In this case, the propane does depend on petroleum refining, and therefore, electricity; but the natural gas may not—depending if it is supplied by a local distribution system tied to an interstate pipeline network. There are a number of gas lights available on the market.

Argand Lamp—The Argand lamp was invented in 1780 by Aime Argand. The Argand lamp had a sleeve-shaped wick mounted so that air could pass through the center of the wick and also around the outside of the wick. The Argand lamp used whole oil, olive oil or vegetable oil. It created a bright steady light and yielded less soot than other lamps of the time. Subsequent inventions—the Carcel lamp and Franchot's moderator lamp were improvements on the design.

Heating and Cooking

Wood is the oldest form of energy used for space heating, water heating and cooking, and occasionally for lighting. It is used by those who choose not to connect to the grid as well as those who do. The first home heating systems were used in ancient times by several cultures—notably the Koreans and the Romans. Figure 3-8 illustrates a Roman system referred to as Orata's Hypocast named after Gaius Sergius Orata. In this system, the exhaust heat from the fire circulates under the home and heats the floor which, in turn, heats the home through conductive and radiant means.

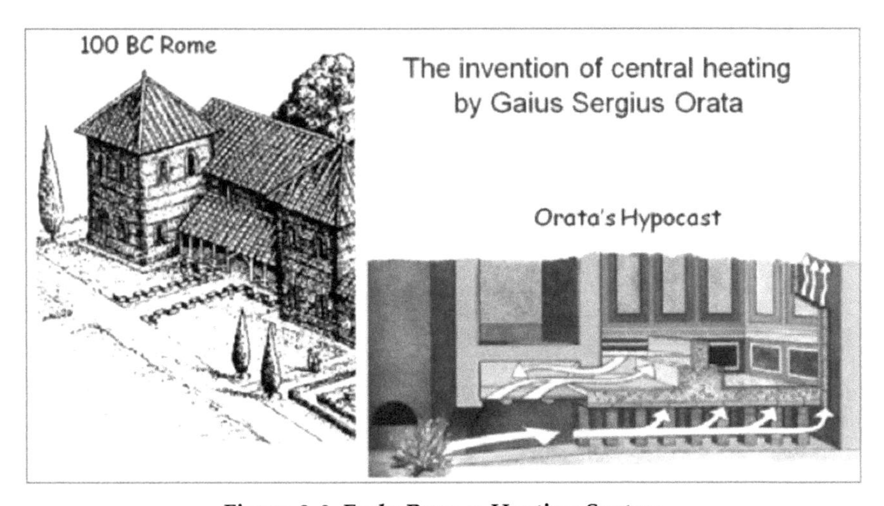

Figure 3-8. Early Roman Heating System

A similar system was used to heat the massive Roman bathhouses. Korean homes, used a similar system as shown in Figure 3-9.

Figure 3-9. Ancient Korean Heating System (Photo by the author)

Today, wood is burned for house heating in more practical wood stove appliances. These can be very effective, and newer models have been designed to burn with limited toxic emissions. Refined wood, in the form of pellets, can now be purchased and used in specially designed stoves. Pellets are torrefied wood—wood which has been heated, partially burned in an air starved atmosphere, and compressed. Pellets allow for easy handling and offer increased efficiency. For cooking and heating water, wood cook stoves are still in widespread use in Amish homes.

<u>Smoke Jack</u>—Even before electricity's wonders were explored, citizens found innovative ways to serve their energy needs. One such example is the smoke jack. The smoke jack is a device which consists of a fan, cog wheel and chain inserted into a chimney flue. The rising heat from the wood fire below turned the fan. The fan, in turn, rotated a wheel and chain which rotated

a spit below just above the fire. Meat was mounted on the spit for meal preparation.

Wood- and coal-burning stoves are still the most widely used for heating and cooking by the Amish. However, kerosene stoves, often called oil stoves, are also used as are propane-fueled stoves. Gas—either natural gas or propane—is often used for home heating, water heating and cooking. Stoves using small portable propane cylinders, called "camp stoves" are also frequently used. These stoves generally have one or two burners.

Solar Heating—The sun is a valuable heat source. A variety of active and passive systems are used to harness the sun in Amish homes for space heating, agriculture (greenhouse), cooking (solar cook stoves), water heating (solar collectors with convection circulation).

Food Preservation

Food preservation can be accomplished without electricity and without use of alternative refrigeration. Food preservation has permeated every culture since the early development of humankind. These preservation techniques usually involve the ability to harness nature, from freezing in the arctic to drying in tropical climates. Food begins to spoil the moment it is harvested. Food preservation enabled man to take root in one place and establish communities. Food no longer had to be consumed or harvested immediately, some could be preserved for later.

Drying—The sun and wind can naturally dry foods. Evidence shows that Middle East and Oriental cultures actively dried foods as early as 12,000 BC in the hot sun. Later cultures left more evidence, and each would have methods and materials to reflect their food supplies—fish, wild game, domestic animals, etc. (Number 2002)

Freezing—Freezing was an obvious preservation method to the appropriate climates. Any geographic area that had freezing temperatures for even part of a year made use of the temperature to preserve foods. Less than freezing temperatures were used to prolong storage times. Cellars, caves and cool streams were put to good use for that purpose. Northern Amish communities have the opportunity to harvest blocks of ice from ponds. Rural ice delivery still exists in several communities.

Fermenting—Fermentation was not invented, but rather discovered. No doubt that the first beer was discovered when a few grains of barley were left in the rain. Fermenting can be done by fermenting fruits into wine, cabbage into kim chi or sauerkraut, and so on. The skill of ancient peoples to observe, harness and encourage these fermentations are admirable. Some anthropologists believe that mankind settled down from nomadic wanderers into farmers to grow barley to make beer in roughly 10,000 BC. Beer was nutritious and the alcohol was divine. It was treated as a gift from the gods.

Pickling—Pickling is preserving foods in vinegar (or other acid). Vinegar is produced from starches or sugars fermented first to alcohol and then the alcohol is oxidized by certain bacteria to acetic acid. Wines, beers and ciders are all routinely transformed into vinegars.

Curing—The earliest curing was actually dehydration. Early cultures used salt to help desiccate foods. Salting was common and even culinary by choosing raw salts from different sources (rock salt, sea salt, spiced salt, etc.).

Jam and Jelly—Preservation with the use of honey or sugar was well known to the earliest cultures. Fruits kept in honey were commonplace. In ancient Greece, quince was mixed with honey, dried somewhat and packed tightly into jars. The Romans improved on the method by cooking the quince and honey, producing a solid texture.

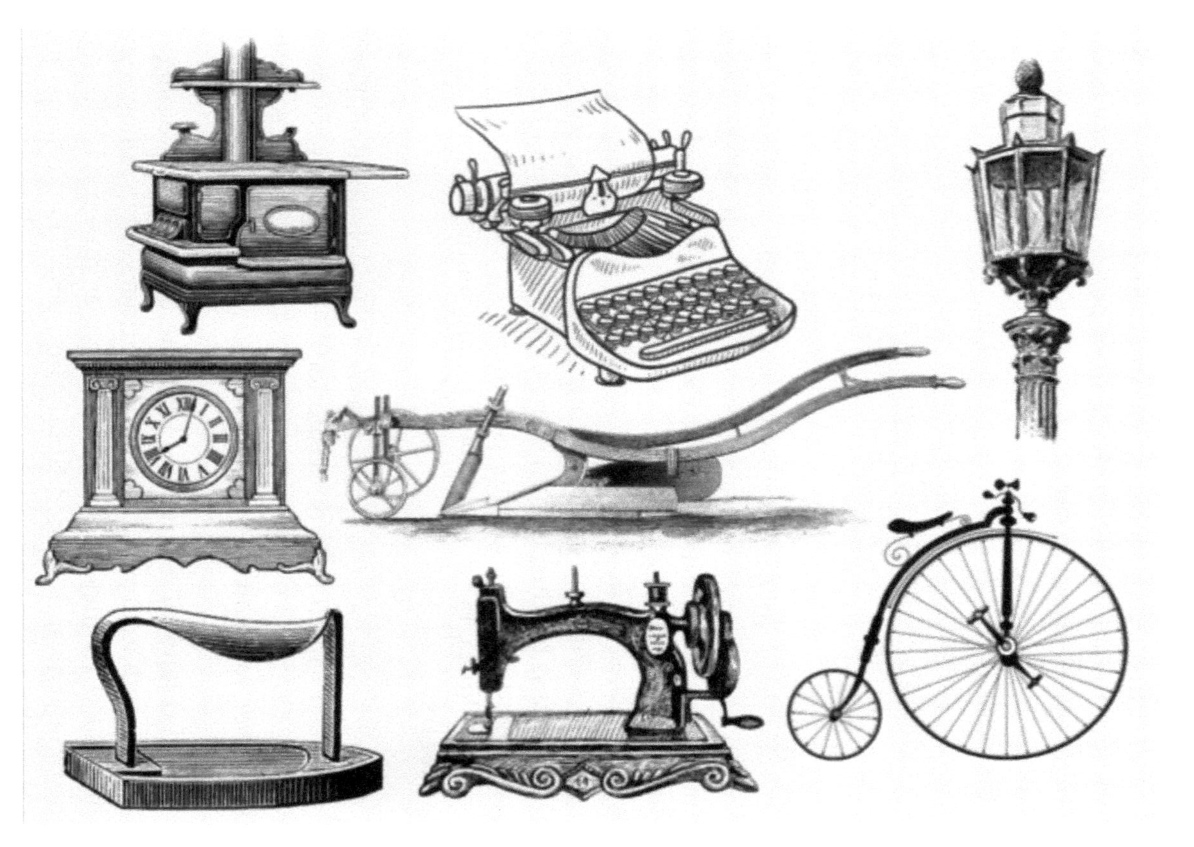

Figure 3-10. Illustrates Various Amish Technologies

Canning—Canning is the process in which foods are placed in jars or cans and heated to a temperature that destroys microorganisms and inactivates enzymes.

Earth Cooling and Root Cellars—An ancient method of cooling has also been used by some cultures—cooling from the earth. Notably employed in desert locations before electricity was available were systems to leverage the cool air found underground to condition buildings. These systems used a series of underground tunnels to supply cold air. Tunnels were dug in long lines radiating out from the structure. Cool air is from these tunnels circulated into the warm structures utilizing natural conventions. Using root cellars is a common technique for ordinary winter storage of fresh, raw, whole vegetables and fruits that have not been processed in order to increase their storage quality. The root cellar is a way to hold these foods for several months after their normal harvest in a cold, rather moist environment that does not allow them to freeze or decompose.

Refrigeration—Some Amish routinely use propane-, natural gas- or kerosene-operated refrigerators. If the Amish home is on a farm and there is a flowing stream present, then the natural chill of the water can be used to preserve food.

Mechanical Refrigeration—Some off-grid communities have adopted the same mechanical refrigeration typically used in recreational vehicles such as camping trailers and motor homes. They can be powered by propane or kerosene. They are sometimes configured to be indirectly dependent on electricity. Often these are alternative systems and also allow operation by 12-volt DC.

Battery-Powered Refrigeration—Refrigerators using the traditional vapor-compression cycle are available to the off-griders which are powered by 12-volt direct current batteries. These are electric devices—but can aid in supporting those who do not have electricity or who wish to divorce themselves from the electric utility.

Water Pumping—Water is often pumped by hand, or with the use of windmills, fossil-fueled internal combustion engine-driven pumps, or waterwheels. Water can also be stored in tanks at heights above the point of use, thereby enabling pressurized or even gravity-flow water systems so as to enable virtually modern plumbing for kitchens and bathrooms. A variety of water pumps are available made from different products including plastic, stainless steel and brass. Some of these pumps can draw water from depths of 200 feet by hand.

Laundry—A popular laundry accessory for those living without electricity is the washboard.

Domestic Hot Water—The Amish have developed several different styles of water heaters which provide hot water for various domestic needs. Washing machines have even been developed with non-electric motor-driven agitators. Today, a wringer washer with a small gasoline engine is found in many Amish homes. Sometimes these are powered by compressed air (Scott 1999), albeit hand-powered machines are still in use.

Clothes Drying—Clothes drying is most often solar-powered by hanging wash on the line. Ironing is facilitated by solid meter irons much as it was in 1850. There are still self-heated irons in use which are either gasoline-powered or where hot charcoal wood be inserted.

Sewing—Tailoring and sewing was and has been by hand.

Other Appliances—The Amish are nothing if they are not inventive. Some have arranged compressors powered by fossil fuels to make compressed air available to kitchens and workshops.

The Financial Impact of Residential Electricity Use

As the proliferation and use of residential appliances increases, it is often assumed that the amount of household expenditures to support that greater use of electricity would increase proportionately. However, that has not been proven to be the case.

The average American household spent 2.82 times as much on telephone service in 2011 than in 1984, but only 2.26 times as much on electricity. In addition, the average household also spent 2.26 times on everything totaled. So over the last three decades, the net value of both telephone services and electricity has rapidly increased. (Bureau of Labor Statistics)

When the first Consumer Expenditure Survey was conducted in 1984, 2.86% of the average American household's expenditures were for electricity. In 2011, the same survey result showed no change in that percentage. (Bureau of Labor Statistics)

As an expenditure, electricity continues to compare very favorably with other household expenditures. Figure 3-11 offers a household daily cost comparison where electricity is paired with other household commodities such as coffee, soft drinks, Cheerios

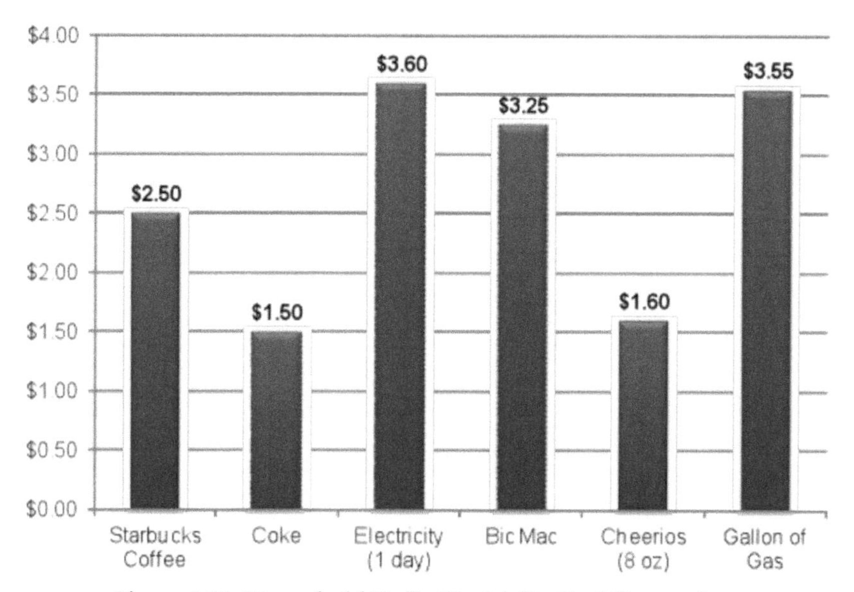

Figure 3-11. Household Daily Electricity Cost Comparison

and gasoline. Electricity remains a bargain, considering the functionality and convenience it affords the consumer.

The value of electricity continues to grow. Based on data from the U.S. Bureau of Labor Statistics, electricity's average annual price increased more slowly than most major consumer goods. This is illustrated in Figure 3-12.

According to the Edison Electric Institute (EEI), while Americans use less electricity today than they have in recent years, as illustrated in Figure 3-13 the portion of household budgets dedicated to electricity bills has declined.

LIVING WITHOUT ELECTRICITY: CONSEQUENCES

There are, however, hidden consequences for living without electricity. For example, the United Nations estimates that when you have no electricity for cooking, heating or light, you burn coal, wood, crop residues, and even dung to meet your basic energy needs—and this causes two million people to die each year due

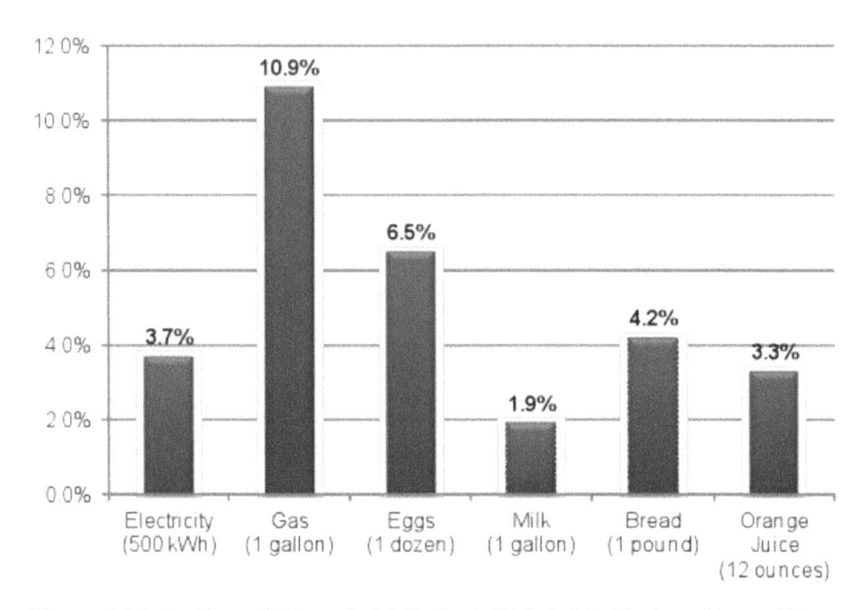

Figure 3-12. Portion of Household Budgets Related to Various Expenditures

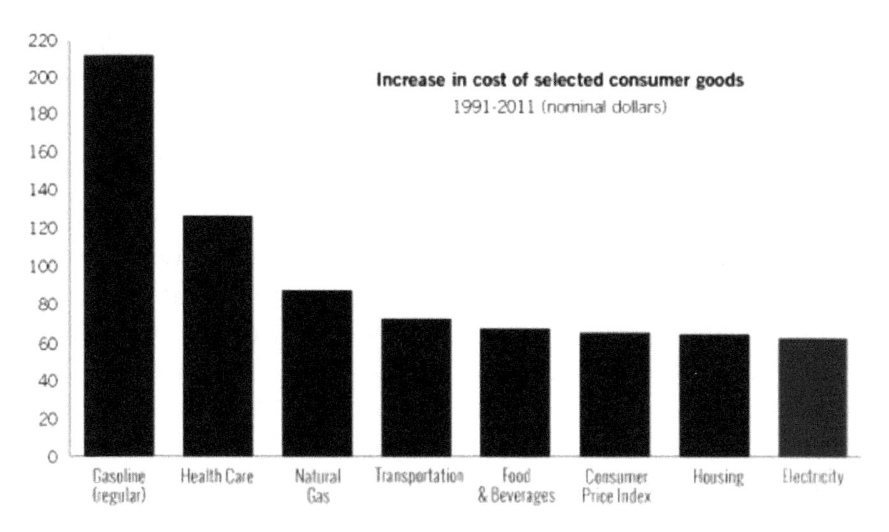

Figure 3-13. Increase in Cost of Consumer Goods (BLS & US DOE)

to inhaling the indoor smoke this burning causes. That's equivalent to 5,400 people who die every day.

Rural electrification has a substantial impact on employment. A relatively recent study from South Africa suggests that by analyzing South Africa's mass roll-out of electricity to rural households and using several new data sources and two different identification strategies (an instrumental variables strategy and a fixed effects approach), researchers found that electrification significantly raises female employment within five years. Electrification appears to increase hours of work for men and women, while reducing female wages and increasing male earnings. Several pieces of evidence suggest that household electrification raises employment by releasing women from home production and enabling micro-enterprises. Migration behavior may also be affected. (Dinkelman 2010)

Domestic technologies like electric appliances are, on the surface, labor saving, but some question if they are actually more enslaving. Some have argued that, based on time studies from the 1920s to the late 1960s, the aggregate time spent on housework remained remarkably constant throughout the period, (Wajeman 2010), albeit there has been some redistribution of time between tasks.

Essential Community Services

Most essential community services are dependent on electricity including municipal water service and wastewater treatment and sanitation services. Clues as to their valve can be gleamed from analyzing the value of clean water from analysis of an interruption in New York City's water supply in 1992. A combination of unbridled development and failing septic systems at the root source of New York City's water system—in upper New York State in the Catskills—began degrading the quality of the water that served Queens, Brooklyn and the other New York City boroughs. In 1992, the U.S. Environmental Protection Agency (EPA) warned that unless water quality improved, it would order the city build a filtration plant.

The plant was estimated to cost between $6 and $8 billion to build and between $350 and $400 million a year to operate. Instead, the city decided to restore the natural health of the Catskill watershed spending $1.3 billion in the process. (NWF 2005) In addition, it is estimated that 1.5 million people per year fish in New York State at a value of $2 billion per year while supporting over 17,000 jobs and over $160 million in taxes.

According to the United Nations, in 2006 at least 1 billion people—or 70% of the world's population—lacked access to clean water. (Dean 2008) Much of that access is dependent on electricity.

Chapter 4

2020 Without Electricity

The new year of 2020 opened with a bang—literally. The traditional bonfire held in New York's Central Park was as big as ever. When midnight came, bells could be heard ringing from every church and from carriages and diesel-powered automobiles which choked the streets. One could marvel at the scientific advances of the time—even without electricity, there have been marvelous developments which allow effective space conditioning, lighting and engine-driven transportation—albeit these uses are not as optimal as society now actually enjoys.

But wait—without electricity? This is a world where electricity as we know it has never evolved. A combination of business issues, fear of electricity and the lack of acceptance of electricity as an energy source by consumers never allowed modern electric power systems—at any scale—to evolve.

So here we are in 2020, and there is no commercial electric systems—no electricity. Previous chapters attempted to address why electricity did not evolve—what happened? In this chapter, we examine what it is like to live in a "modern" world without electricity, and how needs for energy services could have evolved? This takes us back to the late 1870s—just before electricity became available from commercial electric systems.

A DAY IN THE LIFE OF JOHNNY

Johnny is a man in his mid-20s living in New York City. His day begins as the dawn breaks and Johnny's sleep is interrupted by the clanging of a wind-up alarm clock. Fortunately, it is daylight and he need not fumble with firing up the high-efficiency kerosene lantern. There is no radio to help greet the day and no morning DJ, no weather report, no traffic guidance, no tablets or cell phones to check, no email and no text messages to view. Having never conceived such devices—they are not missed. Johnny's house is warm in the winter thanks to a gas-fired heat pump, and cool in the summer thanks to an engine-driven chiller. However, achieving good temperature control with pneumatic controls (without electrical controls) remains a problem. Domestic hot water supplied by a gas-fired water heater gives Johnny a warm shower.

The kitchen has most basic conveniences which an electric future would have—a cook-top, oven and gas-fired absorption refrigerator, all based on either natural gas, propane or liquid fuel. Missing are microwave ovens, digital clocks, electric toasters and any electric motor-driven devices such as dishwashers, mixers, food processors, and blenders. There are internal combustion-driven engines in use in appliances such as clothes washers, but they are complicated and unreliable.

Transport to work or school for Johnny is either by shoe leather, bicycling or engine-driven conveyances driven by using diesel internal combustion-driven engines. These engines are more complex than those that evolved in the electrical world. In the electrical world, there are internal combustion engines whose combustion systems include powerful electronic sparks and carefully metered fuel injectors, all monitored by sensors and controlled by computers. Instead, the only motive-powered systems are either steam engine-driven or diesel-powered. The diesel-powered engines are not started by electric starters, but either by hand cranks or pneumatic-driven devices.

Although the use of horse-drawn carriages are now nearly nascent, traffic in the city streets still moves less fluidly since mechanical traffic lights (at least the early models) were notoriously unreliable and their timing was hard to coordinate resulting in frequent delays and congestion.

Automobiles

Today's modern automobile is typically laced with electrical and electronic devices containing twenty or more computers or programmable—logic controllers. Even from early days when autos first appeared, they depended largely on electric ignition systems driven either by magnetos or the vehicle's ignition system. The ignition system's primary role is as it remains today to provide the spark needed to ignite the mixture of air and vaporized gasoline which is introduced into the engine's combustion chamber. Diesel engines are often an exception to this as they can be operated at higher pressures so that the air and fuel mixture self-ignites.

Early autos were quick to adopt another electrical device—the starter motor. The first autos were started with a hand crank. These cranks were difficult to use, usually too hard to operate by weaker women, children and older men, and dangerous. Broken limbs and dislocated shoulders frequently resulted from the hand crank used to start automobiles in the early years of their introduction. The availability of electric starters, along with other advances, made the automobile a practical appliance. Roughly at the same time, electric lights were installed.

Over the decades which followed, electricity in automobiles added electric windshield wipers, heater motors, electric-powered windows, and more. Today's cars and trucks may have over a dozen computers on board operating sophisticated fuel injection and ignition systems, time-delayed wipers, speed control, high beams, stereos, Global Positioning Satellite (GPS) systems and more.

Johnny Goes to Work

There are simply a multitude of electrical conveniences that Johnny will never experience in addition to microwave ovens, radios, TV's and other electronics. Hairdryers, electric toothbrushes, electric-powered can openers, ice makers, and a variety of other devices are out of his reach.

It is a good chance that Johnny's apartment was a walk-up—no more than six stories high. In the pre-electric days, there were two technologies which limited the height of buildings: one was the use of steel in construction (due to the lack of electric steel making with arc furnaces), and the other was the electric elevator. DC motor-powered elevators were and are essential for mid-rise and high-rise buildings. Replacing those electric motors with fossil-fueled devices was not practical then and is not today.

Subways

The New York subway system is world renowned for its effectiveness in transporting its citizens and commuters from other New York communities and neighboring states. It is, of course, based on electricity—electric-driven motors. Without electricity, a couple of options would be available: Increased use of buses, the development of a cable car system like that in San Francisco or widespread use of street cars. Any of these options would have aggravated traffic and caused more pollution.

Table 4-1 lists the conditions of transportation and energy in New York City in the 1890s.

In Johnny's world, being out at night is not as safe as it is today—street lighting, store fronts and the like are not nearly as brightly lit and illuminated as they are in modern times. Gas lights are not nearly as effective as electric-powered lights and the streets are, therefore, not as safe.

Battery development was stimulated by the development of electrical end use and the electrical distribution system itself. If the electric infrastructure never evolved, then it is fair to speculate that the battery as we know it now did not evolve either. As a result, widely popular portable devices such as flashlights would not exist. Neither would portable electronic devices—a moot point since communications would not have evolved beyond the telegraph.

The transition which western society underwent after the 1870s was shaped by a variety of forces. While some of these are directly impacted by electricity, there were many other factors. To get a complete picture of Johnny's life it is helpful to review life as it was in 1880.

Post-Civil War New York, New York, was the nation's most modern city. The city was made up of low structures. Only church

Figure 4-1. 1886 Lithograph of a busy scene on Broadway in New York City entitled "A Glimpse of New York's Dry Goods District. (Library of Congress).

Table 4-1. Transportation Energy in New York City in 1890—on any give day (Kyvig, 1929-1940)

Transportation Energy	Individual Energy
40 Dead Horses	2,200 Calories Consumed (800,000 per year)
2.5 Million Pounds of Manure	88,000,000 Calories per year used in fuel energy
60,000 Gallons of Urine	

steeples punctuated the skyline. In 1880, the most common jobs were in the factories, on railroads and in railroad stations. There were also a lot of longshoremen and crews on merchant ships as well as miners and loggers. Everything was mechanized, less efficient than it is today and more labor intensive (without electricity). There were telegraph operators and telegraph support people (the batteries were bulky and needed frequent replacement). There were more brick layers and brick makers, barge workers mule wranglers, horse farriers, saddle makers and street cleaners. And there were those who cut ice from New England lakes and transported to city consumers for their use in "ice boxes."

Health Care

1880 health care for Johnny and his family was marginal. It was not until the late 19th century did medicine have a sufficient grounding in science to allow systematic treatment of disease. In fact, in 1880 the US entered a mortality transition when there began a substantial decline in the mortality rate. While fertility levels began to fall in the early 1800s, mortality levels did not consistently decline until the 1880. In fact, life expectancy rose from 40.5 years in 1880 to 60.9 years in 1930. (Fetter) in the 1871-1880 decade there was a life expectancy of about 40 years at birth and at age 40 an expectation of about another 30 years to age 70. (www.jbending.org.uk/stats3.htm)

Value of Electricity: Electricity improves life expectancy
Hypothesis: 50% of the improvement in life expectancy from 61 to 41 due to electricity
Assumptions: 30 year improvement @ 50% / full life value $5.5 Million (2004)
Value of Electricity: $52.80/kWh

America's cities, like Europe, were quite polluted. The major causes of death were water-borne diarrheal diseases. (Meeker, 1972) Engineers in the 1880s embarked on massive public works projects and built reservoirs for filtering water, pipes for distributing it, and sewers to remove liquid waste. (Fetter) Electricity played a key role here.

The 1880 Census was the first census to provide a reasonably complete picture of an individual's health as the day of enumeration by the use of codes. These codes covered virtually every common illness known to the medical community and were categorized in groups such as: infectious disease (e.g. malaria, small pox); chronic disease (e.g. diabetes); and disease of the digestive systems (e.g. ulcers). (www.hist.umn.edu). "Epidemics (typhus, cholera, diphtheria and tuberculosis) were rampant in the city's slums, hiding in the rookeries. Horse manure and human wastes were in the streets. In winter, when all [the street] grime froze, walking on the sidewalks was [almost] impossible. Animals and livestock such as pigs and horses died and remained on the street. "(History of New York City 1855-97, Wikepedia.org)

The development of vaccines and their mass-production, facilitated by the availability of electricity, contributed substantially to human health in the decades that followed.

The Workweek

In the 19th century, the New York workweeks typically ran from Monday through Saturday. This made Sunday the city's day of leisure when the mood of New York drastically changed. "On

Sunday morning New York puts on its holiday dress. The stores are closed, the streets have a deserted aspect, for the crowds of vehicles, animals and human beings that fill them on other days are absent." (McCabe, 1873)

Not all elements of the city in 1873 were so gentile. Many women could be found on Broadway or in the area without male escorts after dark. These women were prostitutes known as street walkers and would often likely be decoys for male thieves. A common scam is for a street walker to lure a tourist to a room and rob him with the help of male accomplices. (McCabe, 1873)

According to some writers, 19th century workers imbibed at incredible levels. Upscale haunts included large saloons and bar-rooms, while the working class was more likely to frequent "Broadway palaces," "gin mills" and the "bucket shops of the five points." (New York in 1872)

Women were not immune to these habits either. As one author started: "wives and mothers, and even young girls, who are ashamed to drink at home, go to these fashionable restaurants for their liquor." (McCabe, 1873)

Perhaps these conditions are nicely summed up by a blogger responding to a Yahoo question: What was city life like in the 1880s? Louise C. responded in 2012:

"The streets would be full of horse drawn vehicles rather than motor vehicles, and there were people who went around cleaning up the horse manure on the streets. Bicycles were becoming popular by this time, they were ridden by men and women. Everyone wore hats when they went out, and gentlemen would raise their hats to ladies they met in the streets. There were some department stores but also many more small shops than there are nowadays. It was common for shops to deliver. They had delivery boys who would take the goods ordered to people's houses. Milk was usually delivered every day, people would provide their own receptacles for the milk. Daily delivery was necessary because without refrigeration, milk went off very quickly. Street entertainers were quite common, there were singers, jugglers, acrobats, etc. who would perform for the passersby. There were many theatres and music halls, dance halls and lots of restaurants, bars, etc. There were also brothels, most cities had a large number of prostitutes. There was a lot of poverty, many people work for low wages, and the poor would live in cramped tenement buildings. Those who were comfortably off generally employed servants. Being a servant was one of the commonest ways for poor women to earn a living." (answers.yahoo.com)

On 1880 the average household size consisted of 4.91 persons and there were a total of 9.8 million US households. Over 80% of households were occupied by married couples. (Gibson, 2012)

Divorce rates in the period of 1880-1886 was 4%. This low rate in the 1800s was due in part to a stigma attached to divorcing a spouse. Women, outside of marriage, had very few economic opportunities. One could not obtain a divorce without proving significant cause of abuse, adultery or abandonment. For a multitude of reasons, the divorce rate by 1985 climbed to 50%. (Jones, undated)

Rural Life Circa 1880

Both the urban and rural population grew rapidly in the 1880s. However, electric technology was not widely adopted in the rural setting until the 1900s. New machines for farming were introduced during this period while people and animals provided most of the power that propelled the machinery. (www.loc.gov/teaches) In 1880 the total US population was 50.2 million with 22.0 million living on farms. There were 4 million farms averaging 134 acres each. (www.agclassroom.org)

Farming during this period was largely a family affair, with family members all sharing a piece of the chores. Women Assumed roles ranging from full partner to manual laborer and performer of menial tasks. (West Virginia History, 1990) Farm houses were quite large with abundant kitchen areas, dining rooms, a sitting room and bedrooms. Women in the home were responsible for: cleaning; washing; ironing; sewing; and meal preparation. These tasks occupied them full time. In 1880 none of these chores were aided by electricity.

Electricity was not prevalent in most farm houses until the late 1940s. Many homes used wood for cooking, water heating and lighting. Increasingly rural residences began to install running water in their kitchens. Laundry was typically done on a washboard and hand wrung. Flat irons were used for ironing. Kitchens lacked refrigeration, but often had ice boxes and stored perishable food in their cellars. Life without electricity on the farm can best be summed up by a quotation from a woman who lived on a farm at the point when electricity first became available: "If I had electricity, I would have everything I need and want in this world." (Eagan, 1990)

The following speculates as to the alternative for humans without electricity:

SO, WHAT IS THE ALTERNATIVE FOR HUMANS WITHOUT ELECTRICITY? LIVING WITHOUT?

- There are no advanced medical diagnostics ➔ Reduced quality of life
- There is limited medical treatment ➔ Reduced longevity
- There is no electronic communications ➔ Reduced productivity, reduced communications, and lack of strong social network
- There is greatly reduced computational ability ➔ Reduced productivity
- Conveniences such as hair dryers/dishwashers/food processors do not exist ➔ Reduced convenience
- There is no microwave cooking ➔ Reduced convenience, increased health risks from food pathogens
- There is no microwave (MW) processing ➔ Less productivity
- There is no electric transportation ➔ Limited and less reliable fossil fueled transportation, if available at all

- There is no ultraviolet (UV) curing ➔ Less productivity
- There is no electric transportation ➔ Limited and less reliable fossil-fueled transportation, if available at all

Electric Process		Fossil Process	Consequence
Electric Infrared (Process Heating & Space Conditioning)	—	Gas Infrared (Process Heating & Space Conditioning)	Inefficient
Electric Motor-Driven Vapor Compression Cooling	—	• Absorption Cooling • Fossil-Driven Compression Cooling	Inefficient, Often Ineffective
Chlorine for Water Treatment	—	Filtration for Water Treatment	Costly, Less Effective
Food Concentration	—	• Gas-Driven Freeze Concentration • Evaporative	Poorer-Quality Produce, Less Efficient
Electric Heat Pump	—	Gas Heat Pump	Less Efficient, Lower Comfort
Electric Resistance	—	Direct-Fired Gas Heating	Environmental Impact, Higher Cost
Plasma Arc Torch	—	Direct-Fired Gas Flame	Environmental Impact, Higher Cost, Lower Productivity

Figure 4-2. Speculates as to which fossil-fueled process may have evolved to substitute for electrical processes and the resultant consequence

There are a number of other consequences which would arise if electricity were not available; the following would result in whole or in part:

- Increased overall use of energy: fossil-fueled end uses are less efficient than electric end uses in most applications.
- Environmental impact of increased fossil fuel use: more emissions such as CO_2, SO_2, NO_x, etc.
- Loss in productivity: many electric industrial processes are more efficient and produce higher-quality products.
- Loss in economic development: overall increase in costs and a decrease in product quality.

- Clean water supply and water or wastewater treatment are more difficult if electric-based technologies are not available. If Johnny indeed lived in New York City, he had the opportunity to be served by a water supply system that was supplied without electricity—from reservoirs in upstate New York.
- Less optimal communication: no digital infrastructure and entertainment limited to local in-person experiences.

Chapter 5

Electricity's Value to Society

DETERMINING THE VALUE OF ELECTRICITY

There are various analytical methods for determining the value of electricity which will be explored in the sections which follow.

Traditional Economists' View of Value

Figure 5-1 illustrates a traditional economist's view of the relationship of supply vs. demand for a commodity as it may be applied to electricity. It suggests that the price of a given quantity is determined by the equilibrium between supply and demand. As consumers demand more electricity, they will be theoretically willing to purchase more until the price exceeds the value they are willing to pay.

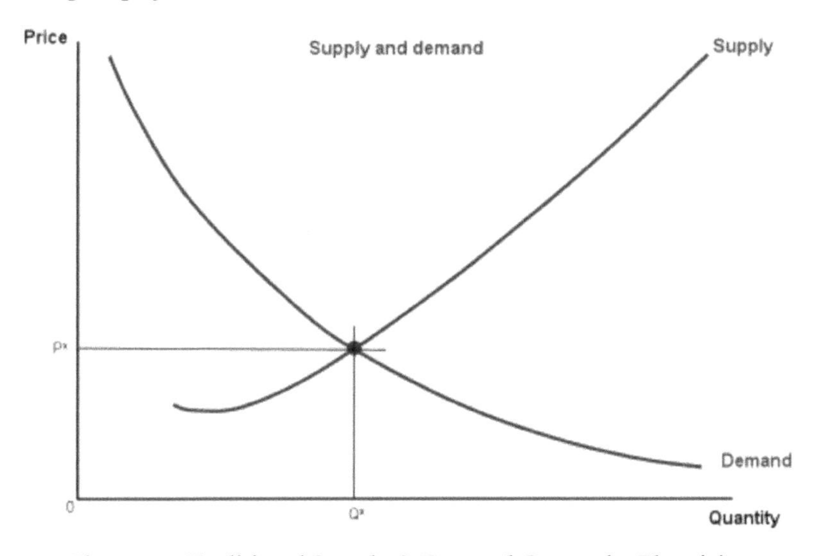

Figure 5-1. Traditional Supply & Demand Curves for Electricity

This is further illustrated in Figure 5-2. Consumers decide to buy the next increment of kilowatt-hours by evaluating either subjectively or analytically the price of that next increment.

It can be inferred that the marginal benefit is equal to the value of electricity. But this does not hold true for commodities like electricity. Electricity is not truly elastic in its demand nor does it respond freely in a regulated market. So its value must be inferred by other means.

Empirical studies typically treat electricity as one of many goods consumed by a household. Strictly speaking this is correct—households do consume electricity. But the demand for electricity is a derived demand and is essentially an input to the production of services from a stock of electricity—consuming equipment in the household. Therefore, it can be argued that electricity is not a part of a household's utility function. Rather, it should be expected that it enters indirectly through the user cost

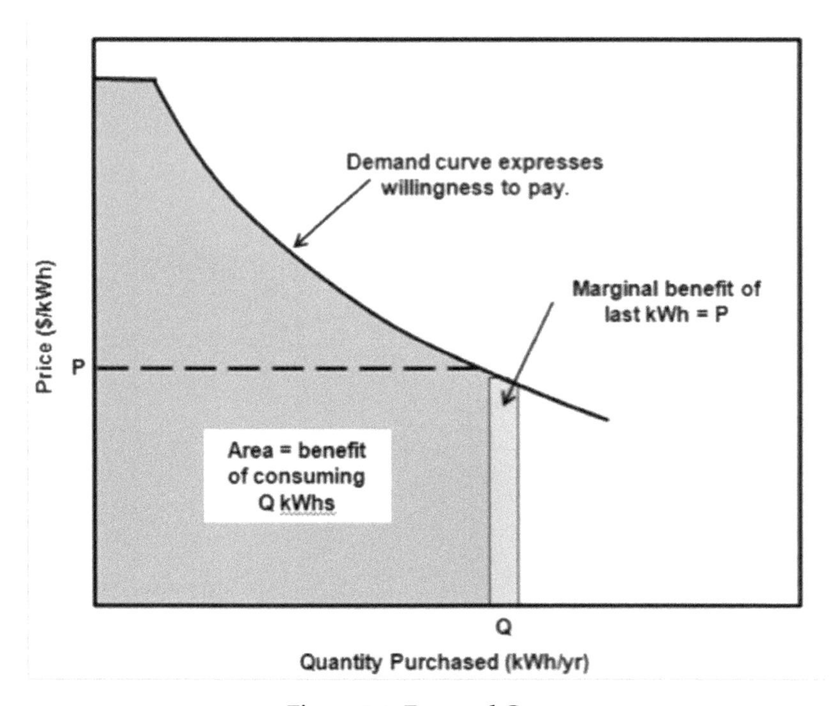

Figure 5-2. Demand Curve

with the services provided by the electricity—consuming equipment.

While everyone instinctively understands that flipping a switch can deliver instantaneous light, electricity's value to society extends far beyond artificial illumination. Electricity has had profound impacts on the world's geographic, technological, and economic development. Early central generating station and electric distribution technology limited the spread of businesses; these businesses had to remain close to power stations because there was no way to "transport" electricity. However, technology advances in both electric transmission and in electric distribution systems as well as advances in power station design facilitated the growth of industry in cities all around the world.

There are increasing numbers of examples of value derived from advanced electric technologies. In medical diagnostics, there are thousands of lives saved each year since imaging such as MRI, X-ray, CAT scan, and ultrasound allow medical professionals to see "inside" the body. Radiation therapy can be used to "kill" cancer cells. Electricity can be used for transportation and off-road applications with reduced overall CO_2 and other emissions. This results in decreased oil imports and enhanced national security. Communications and computations are enabled. Productivity is enhanced substantially allowing GDP to advance at a greater rate than the growth of electricity sales.

Electricity is the solution! If we reinforce and modernize today's grid, increase our use of low carbon energy sources, like renewables, and increasingly develop new electric appliances and devices, we can virtually eliminate most CO_2 emissions while enhancing productivity and assuring energy independence.

Electricity is uniquely valuable—it is a refined energy source. Only electricity can make it possible to harness both the wind, the sun, falling water, the energy locked in uranium and use all of those energy forms to energize some portion of the electromagnetic spectrum delivering unparalleled service to consumers.

In some sense, electricity has become an essential community service and somewhat like clean air or water. Traditional economic theory cannot be used to estimate its value. Valuing electricity might be as difficult as retrospectively valuing the U.S. interstate highway system or the many bridges which traverse streams, rivers or canyons across this great land.

In some sense, this increases electricity's value. As Facebook founder and CEO, Mark Zuckerberg, suggested on September 18, 2013, in an interview with *Atlantic* editor, James Burnett, he doesn't care about Facebook being cool, because now its goal is to be a ubiquitous utility. "Maybe electricity was cool when it first came out, but pretty quickly people stopped talking about it because it's not the new thing, the real question you want to track at that point is are fewer people turning on their lights because it's less cool?" No, because it became essential to modern life.

Natural gas is too valuable to burn in a variety of distributed appliances where it emits substantial CO_2. It should be used, instead, as a feedstock in chemical or plastics production or to generate electricity in highly efficient combined-cycle combustion turbine power plants to produce electricity or in advanced fuels cells. In central stations, unlike distributed appliances, CO_2 emissions can largely be captured, transported and sequestered. And fuel cells are inherently clean.

Industrial Value

The introduction of electricity to manufacturing processes revolutionized both factory designs and production methods. By substituting electric motors for complicated belt-driven machine shop systems, a business could physically and technologically reorganize its manufacturing process to achieve greater efficiencies and economic benefits. In the twentieth century, electricity has been used to develop and power some of the most basic technological developments, including the analog computer, the cathode-ray oscillograph, and the impulse voltage generator. This is why the U.S. gross domestic product (GDP) and electricity grew in lock-step for some time. Without electricity, this would not have occurred.

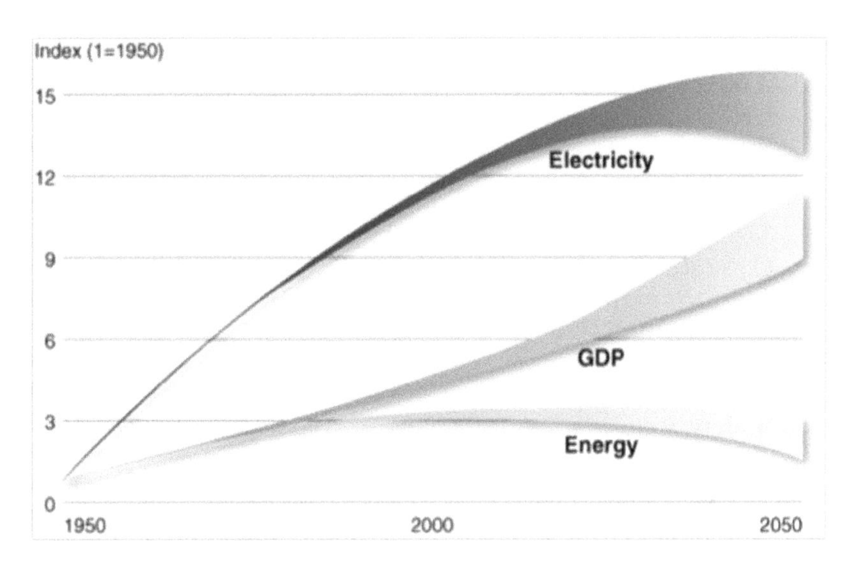

Figure 5-3. Demand Curve Based on Value—How to link GDP with electricity? What is the missing variable?

Electricity is a Uniquely Valuable Form of Energy?

Electricity is a uniquely valuable form of energy, offering unmatched precision and control in application as well as efficiency. It offers unrivaled environmental benefits when compared with other energy options. And finally, electricity provides individuals with a clean comfortable supply of energy. Because of these unique attributes, new electric appliances and devices provide more functionality and require less total resources than comparable natural gas or oil-fired systems.

Electricity's Utility

Electricity's utility is extremely diverse. Certain energy forms can meet one need more efficiently than electricity, but these forms are extremely limited in their range of application. Only one energy form—electricity—can meet all of a customer's energy needs (comfort, convenience, appearance) as well as facilitate the achievement of other needs (medical diagnostics, money from automatic teller machines, personal computers). Electricity is extraordinarily unique in its ability to deliver packages of concentrated, precisely controlled energy and information efficiently to any point.

Electricity is the Solution to the World's Long-term Energy Needs

In addition, electricity can help alleviate many of the concerns facing the world (e.g., environmental problems, limited resources and the spiraling costs for obtaining them). In fact, it is uniquely suited for this critical task:

- It is available from various sources at a reasonable cost.
- Its versatility allows it to be readily converted into easily and efficiently usable forms.
- Its superior efficiency at the point of end use comes from its versatility.
- Affordable, reliable, environmentally sustainable energy for society—only electricity can accomplish this.
- Electricity at the point of end use in most all applications is the cleanest, most efficient, least costly, and potentially most environmentally responsible energy form.
- Applying portions of the electromagnetic spectrum holds limitless potential for improving the quality of life, enhancing productivity, reducing net energy use, and minimizing humankind impact on the environment.
- Only electricity can make effective use of low carbon-emitting energy sources such as wind, photovoltaics, concentrating solar power thermal, nuclear and coal/natural gas equipped with carbon capture and storage.

A recent study by a team of researchers from Tulane University, Carnegie Mellon University, the National Bureau of Economic Research, and the Massachusetts Institute of Technology examined patterns in heat-related deaths between 1900 and 2004. They found that on days where the temperature rose above 90 degrees

Fahrenheit, 600 premature deaths occurred annually between 1960 and 2004—one-sixth as many as would have occurred under pre-1960 conditions (1900 to 1960). (Ellperin 2012)

ELECTRIFICATION

Industry practitioners believe that estimates of the value of electricity can also be inferred from analyzing the value of electrification.

Figure 5-4 illustrates the input that electrification could have on the supply-demand curve.

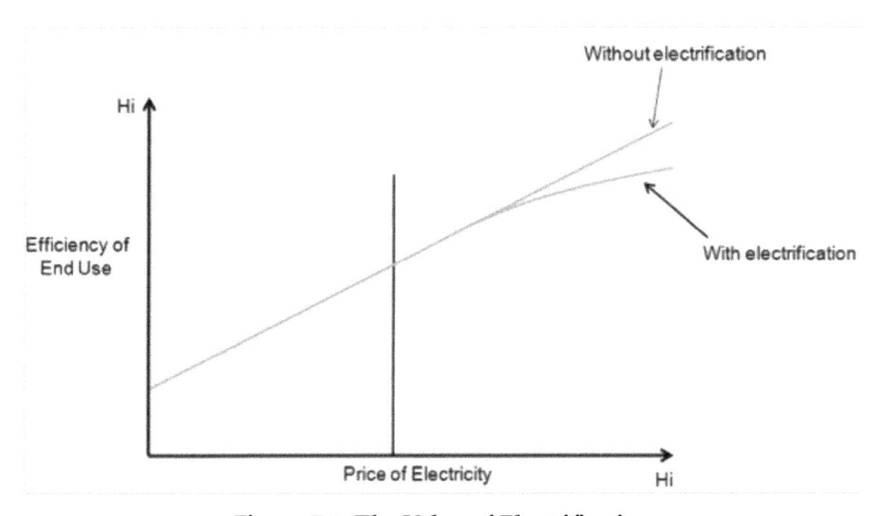

Figure 5-4. The Value of Electrification

Value Based on Substitution

This method provides estimates of value based on the alternative fossil-fueled devices that then evolved as substitutes, is to analyze the individual applications. Figure 5-5 illustrates the range of system efficiency which reflects today's power system. These data reflect very conservative estimates of the efficiency of power systems. The estimate assumes that a fossil fuel is extracted and converted to electricity at an efficiency ranging from 31 to 50%. Note that with increasing amounts of renewable energy, that efficiency could grow 100%. Thereafter, electric power system losses are applied resulting in overall efficiency ranging from 29 to 46%. The comparable fossil system efficiency ranges between 84 and 90%.

Freeze Concentration

Figure 5-6 illustrates a specific application of this template using a comparison of freeze concentration to evaporation. If the electricity enterprise did not evolve, then all the concentration of foods would need to be done with evaporation instead of freeze concentration. The figure illustrates that while the electric process is able to extract between 2.5 and 4.0 pounds of water per 1000 Btu, the fossil-fueled system extract only 1.2 to 1.3 pounds per 1000 Btu.

According to www.dairyfarming.org, the U.S. produces roughly 23 billion gallons of milk annually. If all of the milk was evaporated rather than freeze concentrated, then that would imply a value of electricity of $.0686/kWh.

Value of Electricity: Based on effectiveness of freeze concentration
Hypothesis: Freeze concentrate 20% of milk
Assumptions: 20% of 23 billion gallons concentrated vs. evaporated
Value of Electricity: $.0686/kWh

Electric Heat Pumps

Figure 5-7 illustrates an application involving heat pumps. In the electric case, a heat pump is applied with a coefficient of performance of 3.4. (COP is an overall measure of energy output vs. energy input.)

The United States population based on 2013 data is 304 billion occupying over 117 million households. If every household adopted a heat pump instead of a fossil-fuel furnace, then the implied value of electricity would be $.153/kWh.

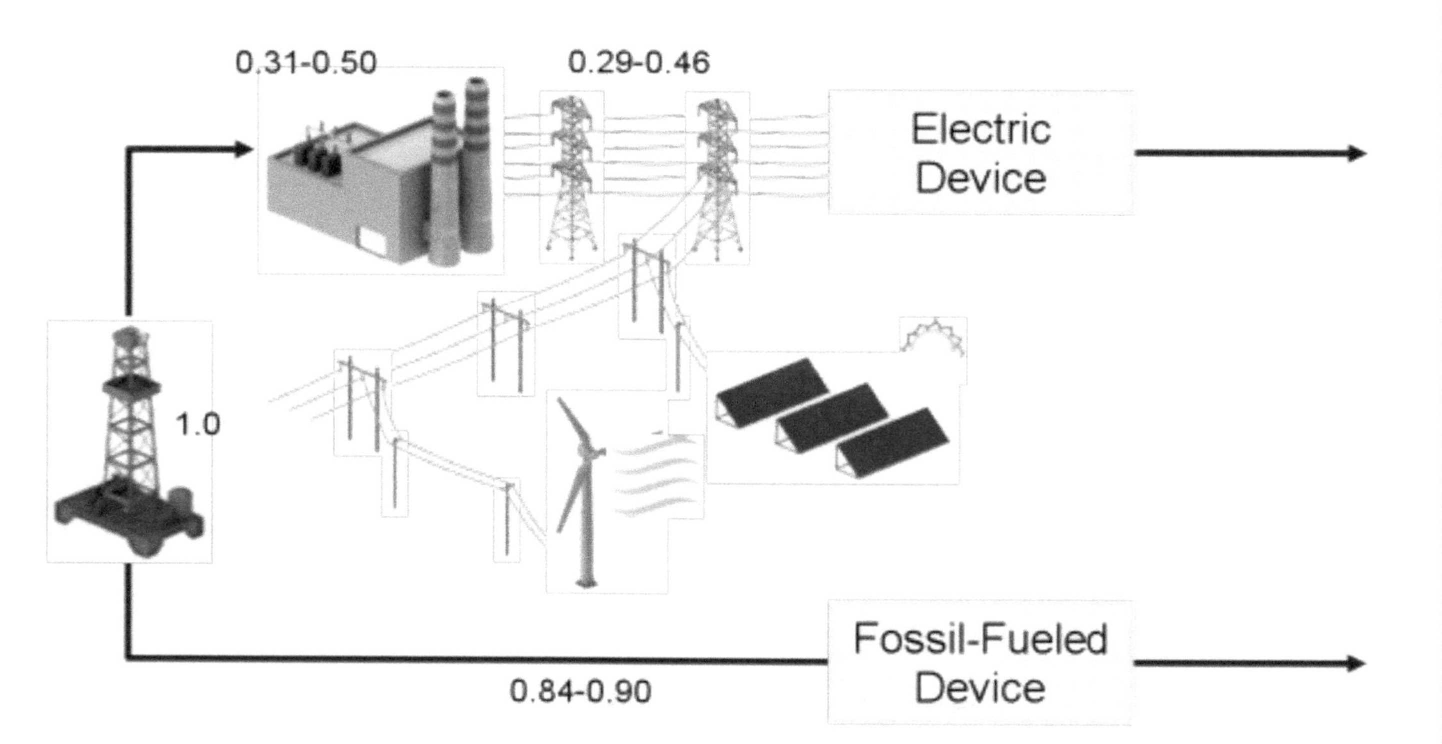

Figure 5-5. Total System Efficiency
*Note that with increasing amounts of renewable energy, that efficiency could grow to 100%.

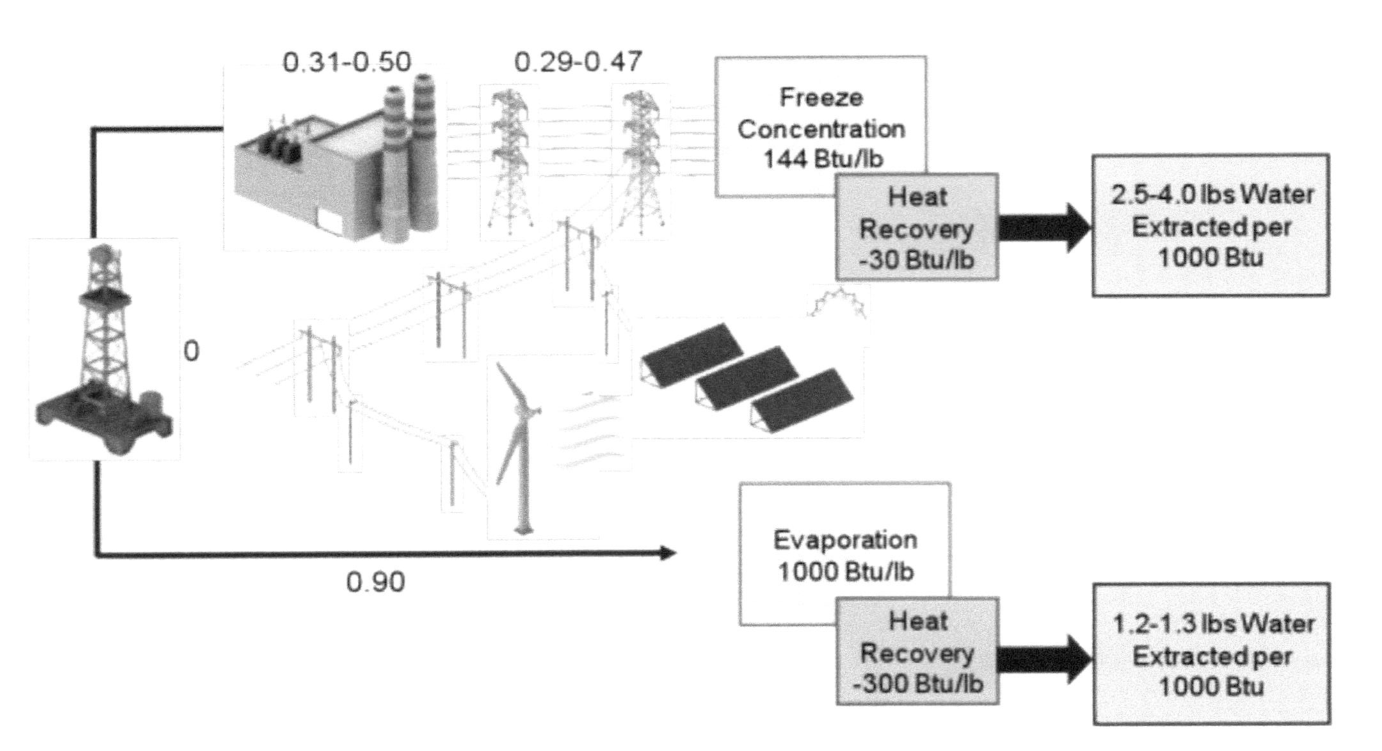

Figure 5-6. Electric Freeze Concentration is More than Twice as Efficient

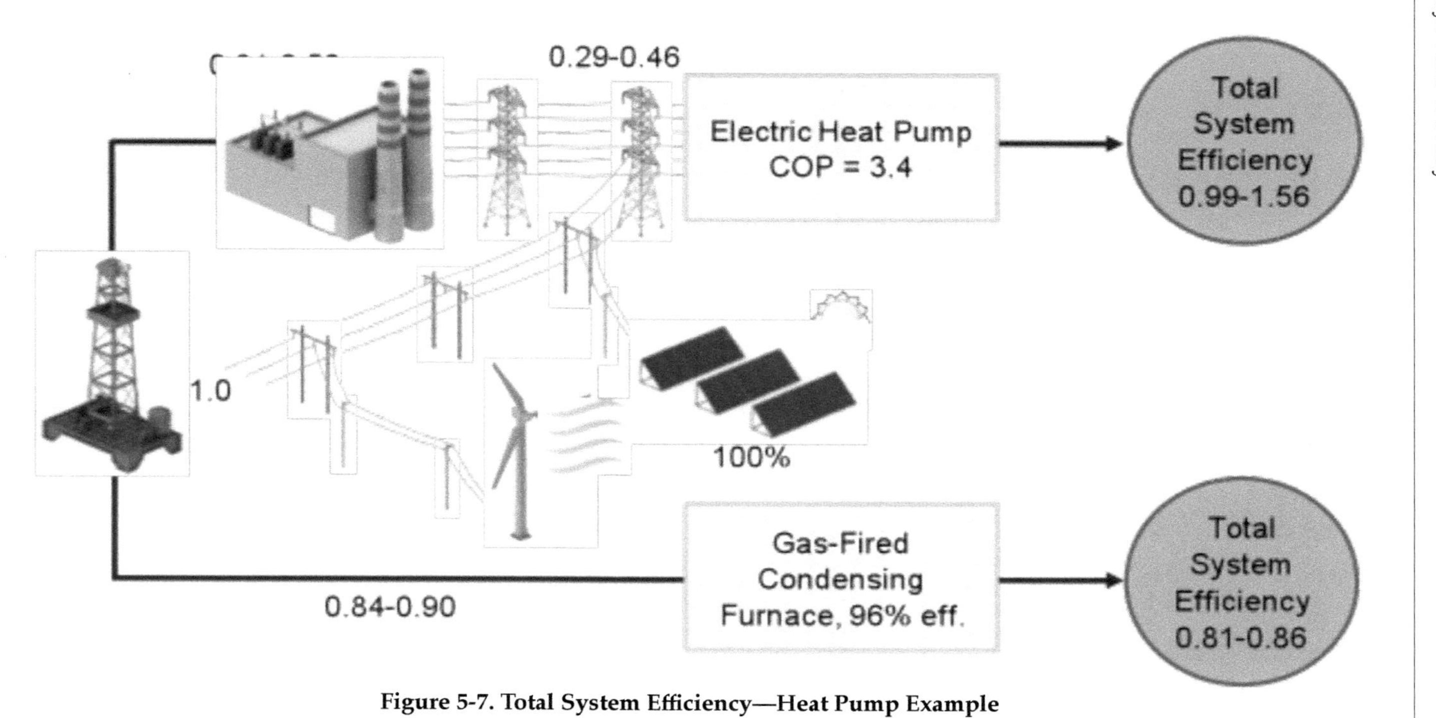

Figure 5-7. Total System Efficiency—Heat Pump Example

Value of Electricity: Based on the effectiveness of electricity in displacing fossil fuels for space heating
Hypothesis: Electricity is substantially more effective in space conditioning implying a value of electricity
Assumptions: 117 million households @ 1.55 efficient and 1500 full load equivalent hours vs. 96% efficient
Value of Electricity: $.153/kWh

Electric Vehicles

According to a 2007 U.S. Department of Transportation (DOT) study, there were approximately 254 million passenger vehicles on the road. (Federal Highway Administration numbers of drivers and vehicles since 1960.) According to the same administration, in September 2013 U.S. drivers traveled 242 billion vehicle miles reaching a cumulative estimate of 2,234 billion for the 12-month period.

If all vehicles were converted to electric, the following would result.

Value of Electricity: As implied from the effectiveness of electric vs. gasoline motor vehicles
Hypothesis: Electric vehicles use substantially less net energy than gasoline powered vehicles
Assumptions: Miles driven 2.234 billion—Electric vehicles equivalent 45 to 72 mpg/gasoline 25 mpg
Value of Electricity: Could save 97 million gallons of gasoline/year = $388/year

Information Transfer

Value of Electricity: Based on electronic information transfer vs. package delivery
Hypothesis: The effectiveness of Information Technology implies that electricity has substantial value
Assumptions: Packages average 14 pages or 16.8 billion pages and need 700,000 barrels of oil $100/bbl/H = $16.67/kWh. Assumes same is displaced by electric
Value of Electricity: Saving $70 Billion/year

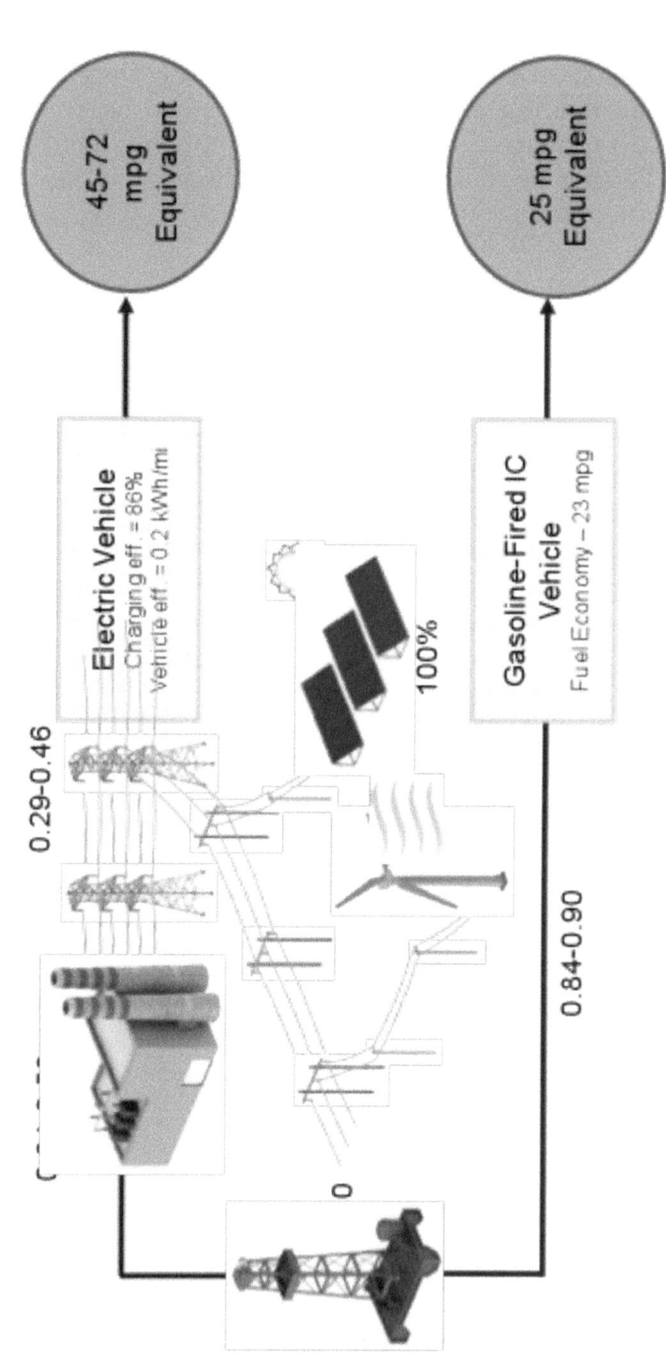

Figure 5-8. Total System Efficiency—Electric Vehicle Example

The advent of digital technology has completely changed society's ability to move information from one place to another. Rather than mailing envelopes or using a courier to deliver a parcel of documents, they can be sent via phone lines or on the internet. Federal Express ships about 1.2 billion packages a year. (www.wiki.answers.com)

Plug-in Electric Vehicles

In one of the most comprehensive environmental assessments of electric transportation conducted prior to 2007, the Electric Power Research Institute (EPRI) and the Natural Resources Defense Council (NRDC) examined the greenhouse gas emissions and air quality impacts of plug-in hybrid electric vehicles (PHEVs). This study can be used to estimate the value of plug-in electric vehicles, in general. The purpose of the study was to evaluate the nationwide environmental impacts of potentially large numbers of PHEVs over a time period leading to 2050. The year 2050 would allow the technology sufficient time to fully penetrate the U.S. vehicle fleet. (EPRI 1015325)

The objectives of the study were the following:

- Understand the impact of widespread PHEV adoption on full fuel-cycle greenhouse gas emissions from the nationwide vehicle fleet.
- Model the impact of a high level of PHEV adoption on nationwide air quality.
- Develop a consistent analysis methodology for scientific determination of the environmental impact of future vehicle technology and electric sector scenarios.

Researchers used detailed models of the U.S. electric and transportation sectors and created a series of scenarios to examine assumed changes leading toward 2050.

- Three scenarios were created representing high, medium, and low levels of both CO_2 and total greenhouse gas (GHG)

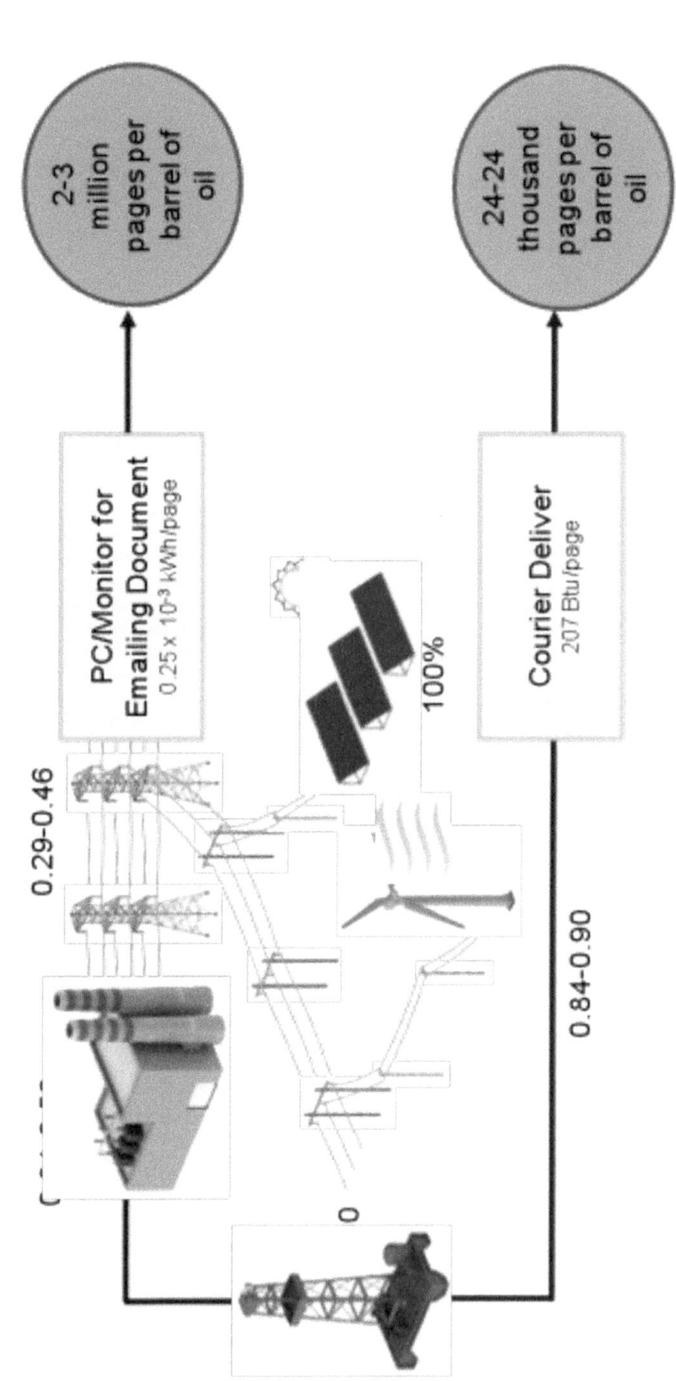

Figure 5-9. Total System Efficiency—Information Transfer Example

emissions intensity for the electric sector as determined by the mix of generating technologies and other factors.

- In each of the scenarios of the emission, three scenarios representing high, medium, and low penetration of PHEVs in the 2010 to 2050 timeframe were overlaid.

From these two sets of scenarios emerge nine different outcomes spanning the potential long-term GHG emissions impacts of PHEVs, as shown in Table 5-1.

Table 5-1. Annual Greenhouse Gas Emissions Reductions from PHEVs in the Year 2050 — (EPRI 1015325)

2050 Annual GHG Reduction (million metric tons)		Electric Sector CO_2 Intensity		
		High	Medium	Low
PHEV Fleet Penetration	Low	163	177	193
	Medium	394	468	478
	High	474	517	612

Researchers drew the following conclusions from the modeling exercises:

- Annual and cumulative GHG emissions are reduced significantly across each of the nine scenario combinations.
- Annual GHG emissions reductions were significant in every scenario combination of the study, reaching a maximum reduction of 612 million metric tons in 2050 (High PHEV fleet penetration, Low electric CO_2 intensity case).
- Cumulative GHG emissions reductions from 2010 to 2050 can range from 3.4 to 10.3 billion metric tons.

Each region of the country will yield reductions in GHG emissions.

To determine the GHG emissions from the electricity generated to charge PHEV batteries, EPRI developed a modeling framework that provides a detailed simulation of the electric sector. The EPRI framework integrates two sophisticated computer models. The first model, the Energy Information Agency's National Energy Modeling System (NEMS) covers the entire U.S. energy-economy system and calculates energy supply and demand nationwide. NEMS outputs—prices and electric loads—are the inputs to the second model, the EPRI National Electric System Simulation Integrated Evaluator (NESSIE). The NESSIE model represents the U.S. electricity sector from 2010 to 2050.

THE VALUE OF EXPANDING END-USE APPLICATIONS OF ELECTRICITY

The use of electricity is generally considered a contributing factor to net carbon dioxide (CO_2) emissions. Growing concern over greenhouse gas emissions has directed research and resources to increasing electric end-use efficiency and to low-carbon power generation technologies to mitigate CO_2 emissions.

There are two main mechanisms for saving energy and reducing CO_2 emissions with electric end-use technologies: 1) upgrading existing electric technologies, processes, and building energy systems; and 2) expanding end-use applications of electricity. Upgrading existing electric end-use technologies embodies replacing or retrofitting older equipment with new, innovative, highly efficient technologies. It also includes improving controls, operations, and maintenance practices and reducing end-use energy needs by improving buildings and processes. In essence, this first mechanism is comprised of what are commonly referred to as energy efficiency and demand response measures. The second mechanism, expanding end-use applications of electricity, involves replacing less efficient fossil-fueled end-use technologies (existing or planned) with more efficient electric end-use technologies. It also encompasses developing new markets for electric end-use technologies that result in overall energy, environmental, and economic benefits.

In 2009, EPRI conducted a study to address the potential for expanding end-use applications of electricity to save energy and reduce CO_2 emissions. The focus was on converting residential, commercial, and industrial equipment and processes—existing or anticipated—from traditional fossil-fueled end-use technologies to more efficient electric technologies. The study began with development of baseline forecasts of energy use and CO_2 emissions. The forecasts were consistent with the US. Department of Energy (DOE) Energy Information Administration's (EIA's) "Reference Case" as presented in its 2008 Annual Energy Outlook (EIA AEO 2008). The study estimates the potential for energy savings and CO_2 emissions reductions during the years 2009 through 2030 for the residential, commercial, and industrial sectors as a function of end-use technology, fuel displaced, and census region. This analysis yielded forecasts of changes in primary energy use and energy-related CO_2 emissions for the U.S. (EPRI 1018906).

Replacing fossil-fueled energy end uses with electric end-use technologies can deliver meaningful net reductions in CO_2 emissions. The most significant reductions can be achieved when implemented in conjunction with low-carbon generation, other efficiency programs, and when specific high-value replacement targets in energy end uses can be identified and acted upon.

Energy Savings

According to EIA AEO 2008, total annual energy consumption for the U.S. in the residential, commercial, industrial, and transportation sectors is estimated at 102.3 quadrillion Btus in 2008, including delivered energy and energy-related losses. The Reference Case forecasts this consumption to increase by 153% to 1180 quadrillion Btus in 2030, an annualized growth rate from 2008 to 2030 of .065%.

The Reference Case already accounts for market-driven efficiency improvements, the impacts of all currently legislated federal appliance standards and building codes (including the Energy Independence and Security Act of 2007) and rulemaking procedures. It is predicated on a relatively flat electricity price forecast in real dollars between 2008 and 2030. It also assumes continued contributions of existing utility- and government-sponsored end-use energy programs established prior to 2008.

Relative to the EIA AEO 2008 Reference Case, this study identified between 1.71 and 5. 32 quadrillion Btus per year of energy savings in 2030 due to expanded end-use applications of electricity. The lower bound represents the Realistic Potential while the upper bound represents the Technical Potential. Therefore, expanded end-use applications of electricity have the potential to reduce the annual growth rate in energy consumption forecasted in EIA AEO 2008 between 2008 and 2030 of 0.65% by 10% to 32%, to an annual growth rate of 0.58% to 0.44%.

The lower bound of these estimated levels of energy savings are potentially achievable through voluntary fuel conversion programs implemented by utilities or similar entities. EPRI's analysis did not assume the enactment of new codes and standards beyond what is already in law. More progressive codes and standards would yield even greater levels of energy savings.

Reductions in CO_2 Emissions

According to EIA AEO 2008, annual energy-related CO_2 emissions for the U.S. in the residential, commercial, industrial, and transportation sectors is estimated at 5,983 million metric tons in 2008. The Reference Case forecasts emissions to increase by 14.5% to 6,850 million metric tons in 2030, an annualized growth rate from 2008 to 2030 of 0.61%. The Reference Case is predicated on a relatively flat CO_2 intensity of the electricity generation mix between 2008 and 2030.

Relative to the EIA AEO 2008 Reference Case, this study identifies between 114 and 320 million metric tons per year of CO_2 emissions reductions in 2030 due to expanded end-use applications of electricity. The lower bound represents the Realistic

Potential while the upper bound represents the Technical Potential. Therefore, expanded end-use applications of electricity have the potential to reduce the annual growth rate in CO_2 emissions forecasted in EIA AEO 2008 between 2008 and 2030 of 0.61% by 12% to 35%, to an annual growth rate of 0.54% to 0.40%.

Figure 5-10 depicts the complete program analysis framework. Steps 1 through 5 in the figure show the steps applicable to potential studies. The focus of the study was on the technical potential; it was beyond the scope of the study to do a thorough evaluation of the effects of economics and customer response and, thus, estimating Economic and Achievable Potentials was not a focus for the study. However, the study considers a Realistic Potential that was estimated using the project team's professional judgment and industry experience. The Realistic Potential is a first approximation to an Achievable Potential. However, a more rigorous analysis using economic and customer preference criteria is needed to estimate a more accurate Achievable Potential.

The study implemented a hybrid top-down and bottom-up approach for determining the potential for saving energy and reducing CO_2 emissions. For each sector, the baseline forecasts for energy use and CO_2 emissions were allocated down to fuel types, census regions, end uses, sub-sectors (for the industrial sector only), and technologies.

The potential of an individual fuel conversion opportunity offered by an efficient electric end-use technology is a function of the opportunity's unit primary energy savings and reduction in CO_2 emissions relative to the fossil-fueled baseline technology. It is also a function of its technical applicability, the turnover rate of installed equipment, and maximum and realistic market penetration values. For a given fuel type and end use, a baseline technology represents a discrete technology choice that complies with minimum existing efficiency standards (to the extent such standards exist) and is generally the most affordable and prevalent technology option in its end-use category. For each end-use category, other technology options are available that use different types of fuels and/or are more efficient. For example, natural gas

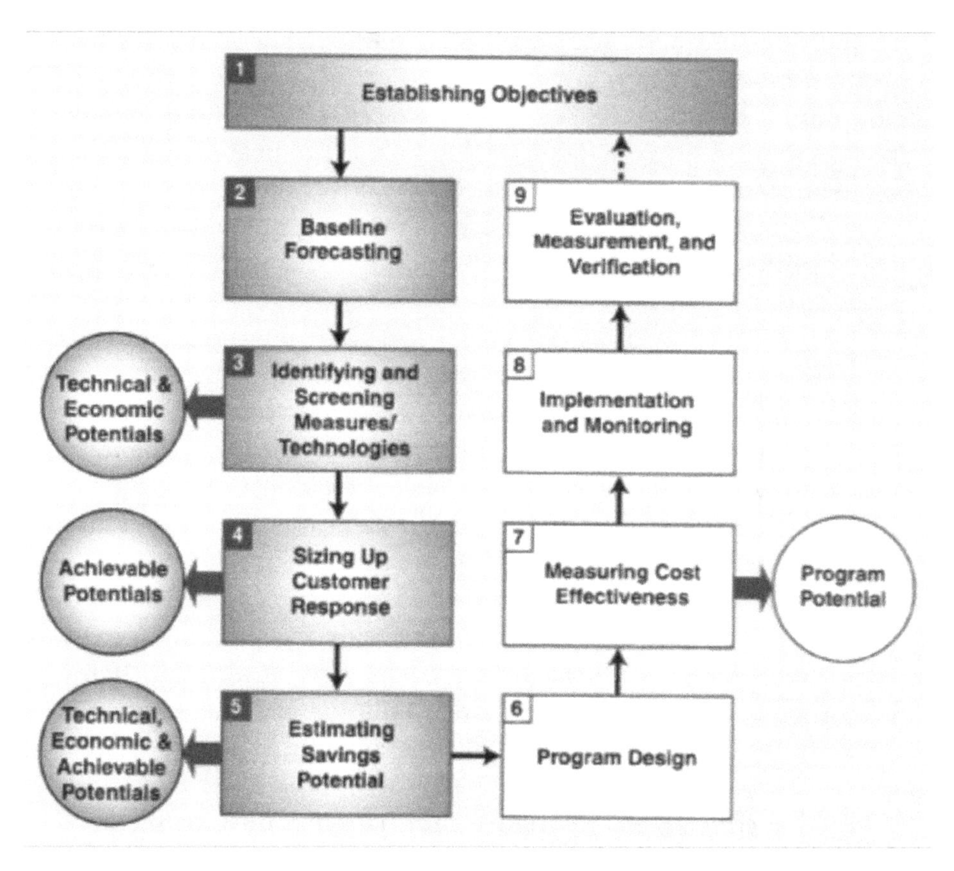

Figure 5-10. Program Analysis Framework (This Study Focused on Technical and "Realistic" Potentials) (EPRI 1016273)

furnaces are a common baseline technology for process heating in the industrial sector. In the approach for this study, efficient electric process heating technologies are applicable in existing industrial applications as replacements for natural gas furnaces that have reached the end of their expected useful life. They are also applicable to future industrial process heating applications anticipated to be met with natural gas furnaces.

The primary focus of this study was to determine the technical potential, which represents the maximum, technically feasible impacts that would result if the selected electric end-use technologies were to displace fossil-fueled technologies. This potential does not take into account cost-effectiveness or customer response, both of which would realistically decrease technology adoption. The study also considers a Realistic Potential. The approach for deriving the Realistic Potential is predicated on first establishing the theoretical constructs of the Technical Potential and then discounting it to reflect market and institutional constraints using professional judgment and industry experience to estimate more realistic values.

This study applies the condition that new equipment is "phased-in" over time. Both the Technical and Realistic Potentials conform to this condition and may be termed "phase-in" potentials. Essentially, the phase-in potentials in this study represent the energy savings and CO_2 reductions achieved if only the portion of the current stock of fossil-fueled equipment that has reached the end of its useful life and is due for turnover is replaced. Thus, the saturation of efficient electric end-use technologies is assumed to grow each year as more of the existing fossil-fueled equipment is up for replacement. In addition, any new equipment being brought on line in a given year due to market growth is assumed to be one of the applicable electric technologies.

Annual Technical and Realistic Potentials by End-Use Sector

Table 5-2 summarizes the Technical and Realistic Potential results by end-use sector for the favorable electric end-use technologies. The results are presented as annual values and thus represent the impacts for the given year—they are essentially a snapshot in time. For the Technical Potential, the table shows that the residential sector has the greatest promise for beneficial impacts. The commercial and industrial sectors follow with values that are roughly comparable to each other. The Technical Potential impacts of all three sectors combined are energy savings of 5.32 quadrillion Btus per year and CO_2 emissions reductions of 320 million metric tons per year in 2030 relative to the baseline forecast.

In the Realistic Potential case, the industrial sector has the highest potential for energy savings, followed by the residential sector and then commercial sector. In regards to the Realistic Potential for CO_2 reductions, the residential sector holds the greatest promise, followed by the industrial sector and then the commercial sector. The Realistic Potential impacts of all three sectors combined are energy savings of 1.71 quadrillion Btus per year and CO_2 emissions reductions of 114 million metric tons per year in 2030 relative to the baseline forecast.

Figures 5-11 and 5-12 graphically display the combined impacts of the three sectors relative to the baseline forecasts for primary energy consumption and CO_2 emissions, respectively. The primary energy baseline data include all delivered energy for the residential, commercial, industrial, and transportation sectors as well as electricity-related losses. Similarly, the CO_2 baseline forecast includes total energy-related CO_2 emissions across all sectors. Both Technical and Realistic Potential impacts are plotted against the baselines.

In terms of the potential for energy savings (Figure 5-11) the Technical Potential is associated with a 4.5% reduction relative to the baseline in the year 2030, while the Realistic Potential yields a 1.5% decrease in 2030. For CO_2 emissions (Figure 5-12) the Technical Potential reduces baseline emissions by 4.7% in the year 2030, and the Realistic Potential reduces baseline emissions by 1.7% during the same year.

These estimates suggest that expanded end-use applications of electricity have the potential to reduce the annual growth rate in energy consumption forecasted in EIA AEO 2008 between 2008 and 2030 of 0.65% by 10% to 32%, to an annual growth rate of 0.58% to 0.44%. In addition, they have the potential to reduce the annual growth rate in CO_2 emissions forecasted in EIA AEO 2008 between 2008 and 2030 of 0.61% by 12% to 35%, to an annual growth rate of 0.54% to 0.40%.

Value of Electricity: In potentially displacing fossil fuels
Hypothesis: New uses of electricity could displace substantial CO_2 emissions by 2030
Assumptions: CO_2 reductions of 114 million tons CO_2 by 2030 @ $30/ton
Value of Electricity: $ 3.42 Billion/year

Baseline	EIA AEO 2008 Baseline Forecast of Primary Energy Use (Quadrillion Btus per Year)			EIA AEO 2008 Baseline Forecast of Energy-Related CO_2 Emissions (Million Metric Tons per Year)		
Sector	**2010**	**2020**	**2030**	**2010**	**2020**	**2030**
Residential	22.2	23.4	25.0	1,259	1,323	1,450
Commercial	18.7	22.0	25.0	1,080	1,265	1,474
Industrial	33.3	34.3	35.0	1,693	1,718	1,733
Transportation	29.0	31.2	33.0	1,980	2,077	2,193
U.S.	**103.3**	**110.8**	**118.0**	**6,012**	**6,383**	**6,850**
Technical Potential	**Decrease in Primary Energy Use (Quadrillion Btus per Year)**			**Decrease in CO_2 Emissions (Million Metric Tons per Year)**		
Sector	**2010**	**2020**	**2030**	**2010**	**2020**	**2030**
Residential	0.352	2.20	2.96	21.7	135	178
Commercial	0.118	0.72	1.09	7.75	47.2	69.1
Industrial	0.123	0.72	1.27	7.30	43.7	73.1
U.S.	**0.593**	**3.64**	**5.32**	**36.7**	**226**	**320**
Realistic Potential	**Decrease in Primary Energy Use (Quadrillion Btus per Year)**			**Decrease in CO_2 Emissions (Million Metric Tons per Year)**		
Sector	**2010**	**2020**	**2030**	**2010**	**2020**	**2030**
Residential	0.055	0.417	0.633	4.97	34.3	47.3
Commercial	0.018	0.246	0.277	1.93	13.3	21.4
Industrial	0.078	0.457	0.802	4.57	27.1	45.1
U.S.	**0.152**	**1.02**	**1.71**	**11.5**	**74.7**	**114**

Table 5-2. Technical and Realistic Potential Annual Impacts on Primary Energy Use and CO_2 Emissions by Sector (EPRI 1016273)

Figures 5-11 and 5-12 graphically depict the impacts of the electric end-use technologies on primary energy use and energy-related CO_2 emissions, respectively. Once again, the values are expressed in terms of the cumulative Technical Potential impacts between 2009 and 2030. In all three sectors, heat pumps are the technology with the greatest promise for saving energy and reducing CO_2 emissions. In the industrial sector, electric arc furnaces have a significant potential for beneficial impacts as well.

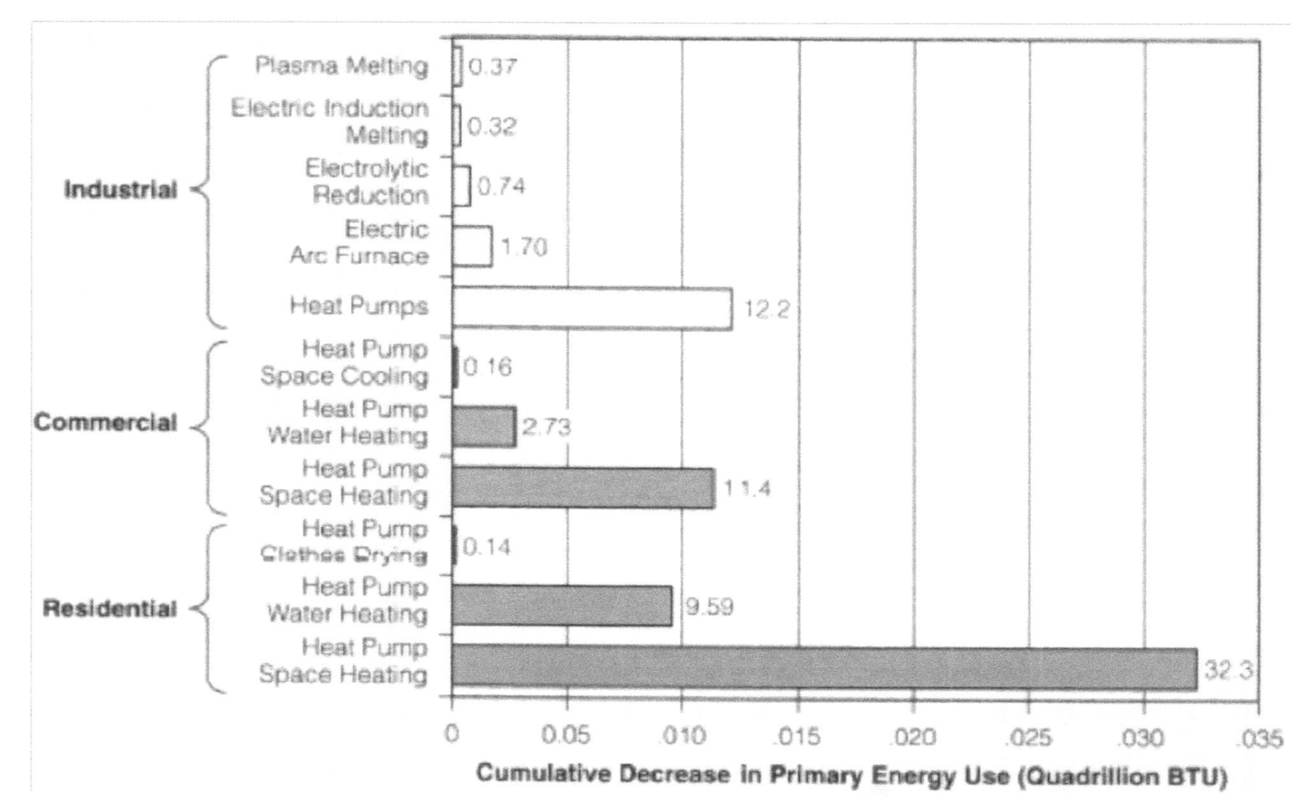

Figure 5-11. Technical Potential: Cumulative Decrease in Primary Energy Use between 2009 and 2030 by Sector and Efficient Electric End-Use Technology (EPRI 1016273)

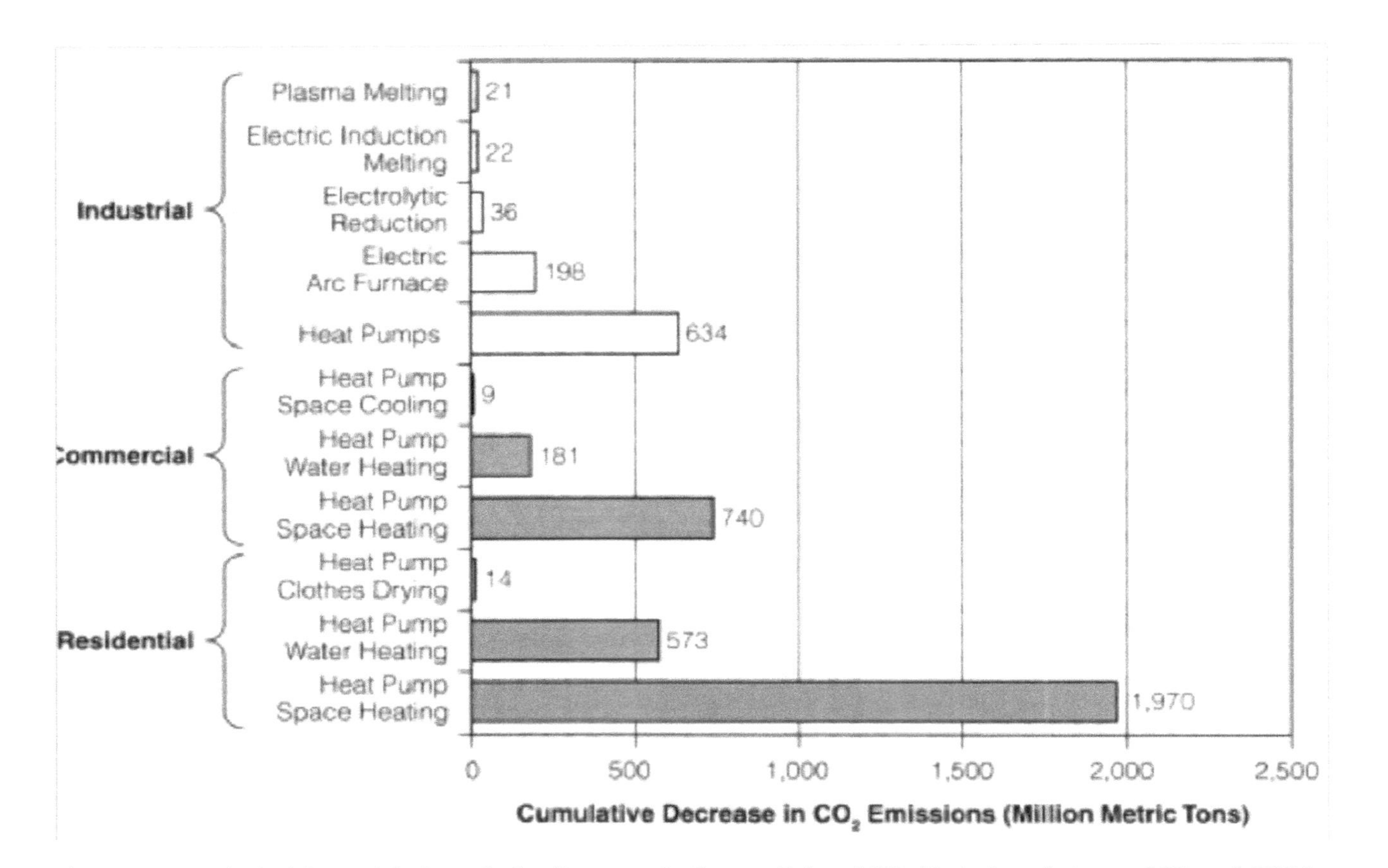

Figure 5-12. Technical Potential: Cumulative Decrease in Energy-Related CO_2 Emissions between 2009 and 2030 by Sector and Efficient Electric End-Use Technology (EPRI 1016273)

In addition, electrolytic reduction, electric induction melting, and plasma melting also show promise. Under a less carbon-intensive future generation mix, more technologies would cross the line and become favorable in regards to saving energy and reducing emissions.

In both the residential and commercial sectors, the end-use areas with the most potential for beneficial impacts are space heating and then water heating. Clothes drying (residential) and space cooling (commercial) also exhibit potential. In the industrial sector, process heating is the predominant end-use area showing potential, followed by space heating.

Value of Selected Technologies

Electricity's unique attributes have enabled a wide range of new technologies which would not have been possible without them. Communications, electronics and medical devices are key among these.

In a recently published study, researchers identified the value of air conditioning. Modern air conditioning can, indeed, be done by either fossil fuels or electricity—but distributing refrigerant or cold air from an air conditioning system is difficult without fans or pumps most ideally powered electrically.

Extremes of heat and cold have a broad and far-reaching set of impacts on the health of individuals. (Adams 2014). Researchers report that a July 1995 heat wave killed 522 people in Chicago alone. Research by the Centers for Disease Control (CDC) found that, on average, 384 people were killed by excessive heat each year during the period 1979 to 1992. (NOAA 1995)

Those at greatest risk of death in heat waves are the elderly without access to an air-conditioned environment for at least part of the day. The issues of prevention and mitigation combine issues of the aging and of public health.

One example of the impact of these new technologies can be examined by observing the economic activity which electricity-related industries simulated. This is summarized in Table 5-4.

Table 5-3. Illustrates the basis for the innovation surrounding these technologies

Electrical Phenomena	Electric Devices	Applications
Electromotive Force Magnetism Motive Power	Motor Generator Electromagnets Magnetic Levitation	(Motive Power) Pumping (Compression) Refrigeration (Including AC & Heat Pumps) Transportation Material Handling Electroforming Maglev Trains
Resistance	Incandescent Light Infrared Heater Cooking Devices Resistance Heater Laundry & Kitchen Appliances	Lighting Space Heating Process Heating Food Service Domestic Hot water
Vapor Discharge	Vapor Discharge Lights Ozone Generator	Lighting Ultraviolet Processing & Curing Germicidal Ozonation Disinfection
Radio frequencies (RF)	Antennas Radio Transmitters & Receivers Wires Metal Detectors Medical Diagnostic Devices Microwave Generators Televisions Telephones RF Heaters	Communications Entertainment Security Processing Medicine
Voltage: Potential Difference & Current	Wires Computers Control Devices Internet Sensors Digital Processing	Communications Commerce Security Entertainment Social Networking

Table 5-4. Economic Activity Stimulated by Electricity—Related Industries—(U.S. Department of Commerce Bureau of Economic Analysis (2011))

	Millions $	% of GDP
Computer and Electronic Products	226,993	1.5
Electrical Equipment	46,798	0.3
Information	646,641	4.3
Computer System Design	198,109	1.3

Enabling Modern Warfare

Electricity has a strong influence on technology used for warfare. Even in World War I (WWI), electricity began to play a role. Weapons were manufactured in electric-powered factories and early forms of communications were deployed, albeit by use of hard-wired landlines.

As World War II (WWII) approached, electricity became much more critical. On one level, it accelerated the severity and complexity of the war effort by enabling the slaughter of millions. And on the other level, it may never have occurred if the advances in technology brought on by electricity were never realized. WWI's and later WWII's advances in electricity brought society modern weapons, radar, submarine, the atomic bomb, depth sounders and many other devices.

Without electricity society would not have been able to wage was as frequently or as ferociously as it had. Using WWII as an example, the many weapons, planes boats, jeeps and munitions including the atom bomb cold not have been conceived or produced without electricity.

Electricity had other impacts on the electric community as indicated in this example from WWII: By December 1942, nearly 200 staff people had left REA (Rural & Electric Cooperatives) for service in WWII. As of 1941, nearly one million farms were receiving power from more than 800 consumer members of rural electric cooperatives financed by Rural & Electric Cooperatives (REA) as nearly 35 percent of all farms were electrified by 1941. (REA 1943). In fact, in 1941 the REA introduced a program called the "Electro-Economy" to stimulate an increase in farm production electrically. The REA staff had convinced the War Production Board (WPB) that electrified farms dramatically pushed up food production. This incentive led WPB to stimulate farms to be electrified.

During WWII, REA instituted the "REA Production Award" to underscore the need to employ electrically powered labor-scarce conditions while meeting war goals.

Medical Technology

One of the most dramatic impacts electricity has had on modern life is in medicine. Today's medical diagnostics, life support and treatment technologies are based on utilizing portions of the electromagnetic spectrum to enable those devices. Should Johnny or his colleagues and friends break a limb, develop a tumor, an abscess, an internal blockage or other internal malady, his/their chances of complete recovery or even survival would be slim without these electrically powered devices.

There seemed to be no end to the early euphoria and strong skepticism about electronic's role in medicine. An article published in the *British Medical Journal* on May 26, 1888, even offered the use of "electricity in gynecology" for the arrest of hemorrhage in cases of ruptured tubal pregnancy. Some practitioners at the time offered strong objections to the concept. (Aveling 1888)

Radiation therapy made possible through the use of electricity has also saved countless lives. Where first introduced in 1950, the *Saturday Evening Post* summarized this phenomenon in an article titled "The Atom May Save Your Life." It told the story of an individual where "The doctors had searched without success for the brain tumor they were sure was causing the patient's severe headache, his nausea, his blurred vision and his staggering gait. The surface of the cerebellum, the region indicated by the symptoms, appeared perfectly normal. Nor had x-rays revealed the tumor's whereabouts. So with great reluctance, the surgeon sewed up the skull and scalp, wrote 'brain tumor not verified' on the chart, and sent the man home."

But the symptoms became worse, and the patient was brought back to the hospital for a second try. This time the medical team enlisted the help of atomic energy. They injected phosphate solution into the patient's veins. The solutions contained radioactive atoms of phosphorus. Then they re-exposed the brain and began to "prospect" for thc hidden tumor with a special needle-thin Geiger counter. The tumor revealed itself.

The miracle of electricity's ability to save lives has endless possibilities. Experiments are underway where patients who are paralyzed are having electrodes and a pulse generator implanted in their spine. Early results are promising, but more work needs to be done. (Waltz 2013)

Indeed the use of electricity in hospitals and clinics illustrates that electricity has become widespread. First, there are the basic functions of a large, complex building. Hospitals have patient rooms, offices, treatment rooms, operating theaters, kitchens, laundries, shops, chapels and lobbies. These require lighting, space conditioning, ventilation and hot water in significant quantities. These facilities likely have pumps, refrigeration for medicine and food, fans, computers and other IT equipment as well. In addition, hospitals now house a variety of extremely critical equipment which diagnoses and treats diseases and illnesses and sustains life.

A stroll through a hospital can likely offer you a glimpse at one or more of the following:

- Life-support machines
- Breathing systems
- Blood gas ampoules monitors
- Dialysis machines
- Magnetic resonance imaging (MRIs)
- CAT (computer-aided tomography) scanning devices
- X-ray machines
- Electrocardiogram (EKG) devices
- Ultrasound devices
- Defibrillation devices
- Electroconvulsive therapy (ECT)

MRIs alone save millions of lives each year. Each year there are over 60 million MRIs done worldwide and 10 million of those in the U.S. alone. Due to the expense of an MRI and the pre-certifications needed for insurance companies, MRIs are conducted on patients with serious conditions. Multiplying these figures for

10 or 20 years, it is easy to conclude millions of lives have been saved, and MRIs are improving the quality of life.

Value of Electricity:
Based of lives saved by modern medical technology
Hypothesis: Use of electricity in modern medicine saves lives
Assumptions: 2 million lives are saved each year by use of advanced, electric—based medical devices
Value of Electricity: Saved or substantially prolonged 40 million lives over 20 years/ value = $248 trillion

The medical field possesses several technologies that save many people's lives, such as heart monitors, defibrillators, x-ray machines, etc. All of these devices are critical to helping the doctors, nurses, and other people in the medical field do their jobs. There are millions of people suffering from diabetes, and because of electric technologies, those people can check their sugar levels and administer insulin by themselves.

Other Life-saving Devices: Ignition Interlock

There are other life-saving applications of electricity that are related to human health. One such device is the ignition interlock device (IID). Ignition interlocks require a motor vehicle operator to blow alcohol-free breath into a device (like a breathalyzer) which is interlocked with a vehicle's ignition before the car can be started and again periodically during travel. They are typically required of individuals who have been convicted of driving under the influence (DUI) of a controlled substance such as alcohol.

According to the Centers for Disease Control (CDC), these devices are effective in reducing repeat drunk driving offenses by 67% (madd.org/interlock), and all offender interlock laws are found to reduce repeat offenses significantly. First-time offenders are serious offenders. Research from the CDC indicates that first-time offenders have driven drunk at least 80 times before they are arrested. According to Mothers Against Drunk Driving (MADD), an interlock is more effective than license suspension alone, as 50 to 75% of convicted drunk drivers continue to drive on a suspended license. MADD data show that states requiring all convicted drunk drivers to use an ignition interlock, such as Arizona, Oregon, New Mexico and Louisiana, have cut DUI deaths by more than 33%.

Value of Electricity: Reducing DUI deaths by ignition interlocks
Hypothesis: Ignition interlocks reduce DUI deaths.
Assumptions: 33% of 10,228 lives saved
Value of Electricity: $23.4 Billion/year

What Other Urgent Needs Could Electricity Fulfill?

One of the most futuristic images of what lies ahead in the future of electricity can be gleamed by studying the phenomenal promises suggested in the TV and movie series "Star Trek." Looking back on the shows, the following are some of the promises which came to life in whole or in part.

Food Replicator—The food replicator depicted in Star Trek may have seemed simply inconceivable, but there are several technology developments which suggest that the fantasy depicted in the media would include a promise of possible reality. For example, Fabs@Home (www.fabsathome.org), an open source mass collaboration, is developing personal fabrication technology aimed at bringing personal fabrication to individuals. Fabs@home tools have the unique ability to make objects out of multiple materials in what is often referred to as 3-D printing.

Teleportation—the "Beam me up, Scotty" feature of Star Trek technology is one of the hardest to conceive. The concept is to allow humans to travel long distances without physically crossing the space between. (http://science.howstuffowrks.com/science)

Transparent Aluminum—Transparent aluminum is reality. It's called ALON and is a transparent aluminum armor made of aluminum oxynitride. (http://ww.howstuffworks.com/transparent-aluminum-armor.htm)

Wrap Drive—Perhaps prompted by the allure of Star Trek's

Warp Drive, NASA has been pursuing options to transport a spacecraft to the nearest star in a matter of weeks. The concept requires space time to be warped both in front and behind a spacecraft. (http://io9.com)

The Holodeck—Another fascinating and maybe among the most unfathomable futuristic technologies displayed on Star Wars episodes is the holodeck. While this technology may never replicate "The Secret Life of Walter Mitty," it does provide an extraordinary goal for scientists. At the University of California students made inroads in creating alternate universes. (http://www.projectholodeck.com).

The Tractor Beam—In the Star Trek series the U.S.S. Enterprise would often tow another spaceship using an electronic beam (a beam of light) called a "tractor beam." A miniature tractor beam has been developed in a collaboration led by the University of St. Andrews, et al. Their tractor beam uses beams of light to move miniature particles. (http://forbes.com)

The Tricorder—In Star Trek, a handheld instrument, referred to as a Tricorder was used as an environmental sensor and a personal medical assistant. The ideal solution is a miniaturized, handheld device which includes medical testing, explosives detection and food safety. Prototype devises have been developed. (http://www.theengineer.co.uk)

Communications

The term Base of the Economic Pyramid, or BOP, is used to describe those countries that are literally at the bottom of the pile economically. These include the countries in Africa, Asia, Eastern Europe and Latin America. One example of the value of Information and Communication Technology (ICT) is presented by the World Resources Institute (WRI). The WRI offers the hypothesis that those in the BOP cannot join the global economy, and benefit from it, until they are connected to it. There is demand for such connections among the BOP and a willingness to pay—because the value proposition, for someone without connectivity, is compelling. Table 5-5 lists their intended budgets for ICT.

Table 5-5. Base of the Economic Pyramid (BOP) Countries Spending on ICT (dollars in billions)

Region	Spending
Africa	4.4
Asia	28.3
Eastern Europe	5.3
Latin America	13.4

There are numerous examples of how electricity-based technology can save lives. A 2011 report out of the National Ocean and Atmospheric Administration (NOAA) shows in stark, dramatic fashion just how crucial technology is to protecting families from harm or even death. (MacDonald 2011). NOAA reports that in 2010, its "satellites were critical in the rescues of 295 people from life-threatening situations throughout the United States and its surrounding waters." The satellites picked up distress signals from emergency beacons carried by downed pilots, shipwrecked boaters and stranded hikers, and relayed the information about their location to the first responders on the ground.

NOAA reports that of the 295 rescues that year, 180 people were rescued from the water, 43 from aviation incidents, and 72 in land situations where they used their handheld personal locator beacons. The signals were picked up by polar-orbiting and geostationary satellites which work along with Russia's COSPAS spacecraft as part of the international Search and Rescue Satellite Aided Tracking System, called COSPAS-SARSAT. This system network of satellites quickly detects and locates distress signals from emergency beacons onboard aircraft and boats, and from smaller, handheld personal locator beacons.

Value of Electricity: NOAA saves lives by use of rescue communications

Hypothesis: Rapid response rescues at see save lives
Assumptions:295 people saved each year
Value of Electricity: $20.4 Billion/year

Cell Phones

Cell phones are certainly among the modern marvels we have derived from electricity. The concept has been around since 1943 when cartoonist Chester Gould sketched the fictional two-way wristwatch/radio on the comic strip character, Dick Tracy. The inventor of the cell phone, Dr. Martin Cooper from Motorola, made the first call in 1973 leading the Federal Communications Commission (FCC) to encourage the cellular idea. Today, mobile technology is changing the way we live. Reportedly, 49% of Americans used smart phones in 2012, and fully four out of five smart phone users check their phones within the first 15 minutes of waking up—and 80% of them say it's the first thing they do in the morning. (Qubein 2013) Indeed, smart phones are one of the electrical devices in the information technology space that have changed the way we live.

Mobile communications, essentially enabled by the electric infrastructure, is by a wide margin the fastest-growing giant industry on the planet. All major digital technologies are headed to mobile—telecoms, computers, the internet, etc.—and all major media are headed to mobile—music, gaming, news, television, and advertising—and even money, from coins to banking to credit cards, is headed to phones. (CVA 2011)

All totaled, there are 5.2 billion far more than those who use personal computers; landline phones; automobiles; television sets; credit cards; the internet; banking accounts; and radio receivers in use worldwide. According to Tomi Ahonen Almanac 2011, the world's population is 6.9 billion representing an active mobile phone subscription for 75% of them.

The services side of mobile is nearing the trillion dollar milestone itself, hitting 928 billion dollars in value in 2010. Mobile advertising revenues have hit $8.8 billion, and mobile apps are worth $9 billion.

Smoke Detectors

A seemingly simple, electricity-enabled technology is the smoke detector. The latest available data from the National Fire

Prevention Association (NFPA) show that fires in the U.S. caused over 3000 deaths, 17,500 injuries, and $11.7 million in property damage in 2011. However, homes that contain working smoke detectors suffer 90% less property loss and 75% fewer fatalities than homes that don't. This makes an obvious case for smoke detectors. (Giaimo 2013)

Value of Electricity: Saves lives by powering smoke detectors
Hypothesis: Smoke detectors are very effective in saving lives and reducing fire damage.
Assumptions: 75% of 3000 deaths could be saved by use of smoke detectors
Value of Electricity: $15.6 Billion/year

Self-driving Car

One of the more innovative electric-based technologies which combines electric-based sensors, communications and computational ability is the self-driving car. (Carlson 2013) Google claims that if the self-driving car is perfected, it could save 1.2 million lives per year.

According to the U.S. census (http://www.census.gov), 38,808 people died in car crashes in the U.S. in 2009, and according to the World Health Organization (http://www.who.int), 1.2 million people die in car crashes around the world each year.

According to news reports, Google's self-driving car program has a primary goal to eliminate the 99% of deaths that are caused by "human error."

The self-driving car uses sensors which can "see" 360 degrees around the car. They remain constantly alert and provide information to on-board computers so there are virtually no blind spots. Cars are already available which warn drivers who follow too closely and even apply brakes when the driver fails to do so. Other sensor systems recently available can parallel park cars.

In 2014 alone, six major automakers released details on self-driving car programs. (*Popular Science,* 2014). Volvo is perhaps the first to put them in consumers' hands. They plan to have

100 cars on the road by 2017. In early September, 2014, Mercedes demonstrated a self-driving "Future Truck 2025" in Stuttgart. The truck sports radical LED lighting, aerodynamic design, and a radar and camera system to help the truck drive itself. (Ross 2014) As Tomas Broberg, senior technical advisor for safety at the Volvo Cars Safety Centre, refers to it, revolution has started with collision avoidance auto braking, steering and autonomous driving. (*Popular Mechanics*, October, 2014) to make cars even smarter, the federal government has been testing vehicle-to-vehicle communications. With it, cars can warn cars about their position.

Saving lives is not the only benefit from having electricity available in cars, but according to *IEEE Spectrum*, smart cars would save 420 million barrels of oil in 10 years. Assuming a value of $100/barrel, that means an additional value of $42 billion over 10 years from electricity. (Jones, 2014). Not burning the petroleum would also prevent 70 million metric tons of CO_2 from being released into the atmosphere.

If there is not electricity—there may be cars—but there would not be self-driving cars.

The revolution of self-driving cars will not end with automation of the car, "robots can already vacuum your house and drive your car" (Piore, 2015). Researchers in Japan are already developing robots that can replace humans in taking on a multitude of tasks—even learning to replicate actions of human. Setting aside their potential to provide social interaction with humans. They are increasingly demonstrating the ability to use electricity to increase the quality of life.

As humans learn to live with robots, their ability to displace humans and to expand our ability to experience greater cultural dimensions and learning experiences could be without bounds.

Chapter 6

Using the Economic Impacts of Blackouts to Estimate Value

WHAT IS A BLACKOUT?

The great Northeast Blackout of 1965 occurred at 5:27 pm on November 9, plunging 30 million people into darkness. At 5:21 pm, New York WABC radio's famous personality, Dan Ingram, told his listeners that electricity was slowing down. It was "slowing"* from 60 hertz to 56 hertz and then to 51 hertz 2 minutes before the blackout took effect. (Mitnick 2013 and WABC Radio).

One method often used to assign a value to electricity is to assess the impacts of blackouts. Blackouts are typically identified as a prolonged loss of electric service from storms or other extraordinary events. Blackouts are often characterized as lasting one hour or longer and impacting 50,000 people or more.

Blackouts or power outages, particularly from large disturbances can cause:

a. Loss of life
b. Loss of productivity
c. Loss of production
d. Regional economic loss
e. Flow-down losses in degraded service from municipal water, sewage, telecom, mass transit, etc.

---

*Electricity doesn't actually "slow." However, in alternating current (AC) systems, the frequency or change in cycles from positive to negative can decrease from a nominal 60 hertz to something less. Once a power system is operating appreciably less than 60 hertz, it can become unstable. So a decrease from 60 to 56 reflects moving to a lower frequency.

The evidence is clear that, in recent years, extreme weather events have severely affected the U.S. economy. Disruptions to our power system from natural events deliver serious consequences in today's technology-driven cultures; nations depends on a reliable, resilient, safe, and secure electric power system to ensure vital necessities such as moving cargo and passengers on transportation systems, operating cellular networks and data centers, running fuel pumps, providing business and consumer

The Miami Herald

WEDNESDAY, SEPTEMBER 29, 2004 | 102ND YEAR, NO. 15 | ©2004 THE MIAMI HERALD | FINAL | 35 CENTS

AFTER THE STORM

FAMILY CIRCLE: Michelle Nedd, left, and son Terelle share light of a tiki torch outside their Fort Pierce home. Her children are learning this week how much they've taken electric power for granted, she says.

LIVING WITHOUT POWER

Figure 6-1. Illustrates the Practical Consequences of Blackout (The Miami Herald)

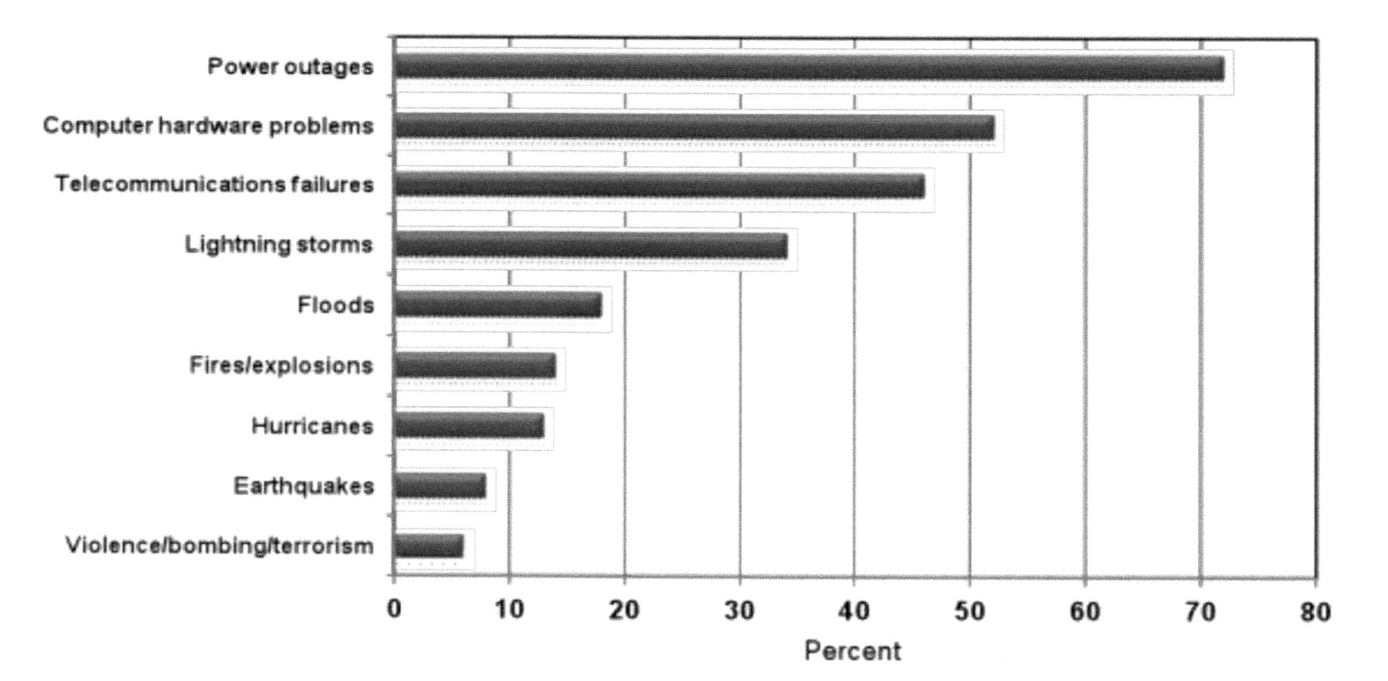

Figure 6-2. Illustrates the various causes of business disruptions.

access to banking systems, and maintaining home climate control and refrigeration. (Gridwise Alliance 2013)

On any given day, the equivalent of 500,000 people in the U.S. are without power for two hours or more.

POWER OUTAGES

Extreme weather is the leading cause of power outages in the United States. Between 2003 and 2012, an estimated 679 widespread power outages occurred due to extreme weather. Power outages close schools, shut down businesses and impede emergency services, costing the economy billions of dollars and disrupting the lives of millions. The resilience of the U.S. grid is a key part of the nation's defense against severe weather and remains an important focus of President Obama's administration. (Executive Office of the President 2013)

Outages caused by severe weather such as thunderstorms, hurricanes and blizzards account for 58% of outages observed since 2002 and 87% of outages affecting 50,000 or more customers. In all, 679 widespread outages occurred between 2003 and 2012 due to severe weather. (U.S. DOE, Form OE-417) Furthermore, the incidence of severe weather is increasing.

The Office of the President estimated the annual cost of power outages caused by severe weather between 2003 and 2012 and described various strategies for modernizing the grid and increasing grid resilience. Over this period, weather-related outages are estimated to have cost the U.S. economy an annual average of $18 billion to $33 billion. Annual costs fluctuate significantly and are greatest in the years of major storms.

Value of Electricity: Cost of weather related power outages
Hypothesis: Weather related power outages cost to U.S. economy give some indication of the value of electricity
Assumptions: Each major weather event costs over $1 billion in power outages
Value of Electricity: Average impact on the U.S. economy is between $18 and $33 billion per year

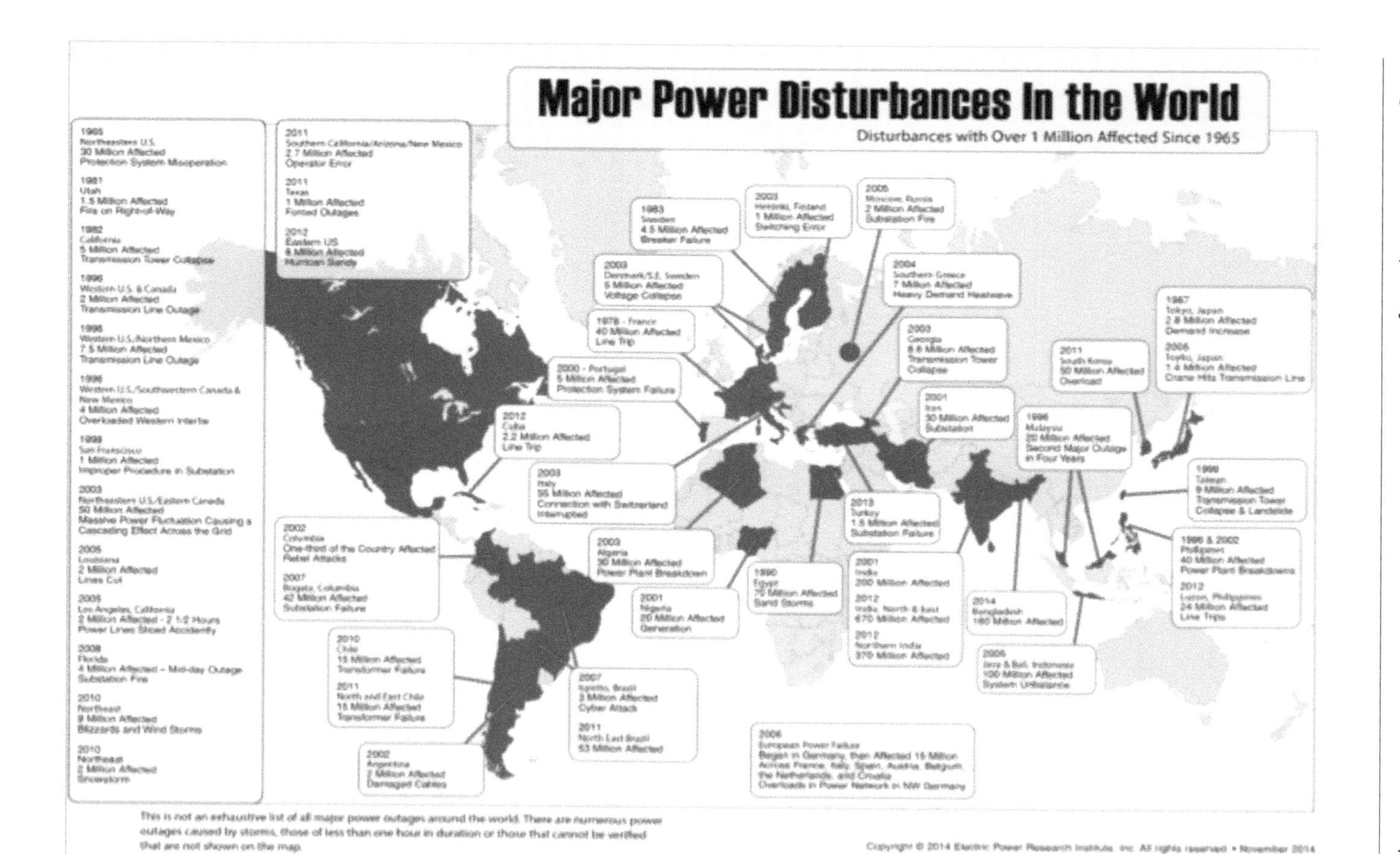

Figure 6-3. Illustrates blackouts around the world (EPRI)

In 2012 alone, there were 11 different weather and climate disaster events across the United States with estimated losses exceeding $1 billion each. Every dollar spent on hazard mitigation saves at least $4 in avoided cost if a disaster strikes again.

The Infrastructure for a Digital Society

A number of studies have highlighted the losses attributed merely to poor power quality (PQ). There are published estimates of $12 billion per year in the U.S. alone. *Fortune Magazine* (12/21/98) stated $10 billion per year. An AEP/Siemens/Bechtel JV press release quoted $26 billion per year in lost production with a single shutdown costing $500,000 per hour.

And the investments to deal with power quality serve as examples. For instance, projections for worldwide uninterruptible power supply equipment: $310.2 billion in 1997 and with a compound annual growth rate of 12.5 reached $629 billion in 2003. The North American market share was 35% in 1997 with Europe and Asia following closely. Overall: 20% of the traditional electricity market is for high-value energy products during this period.

Lawrence Berkeley National Lab Cost of Power Interruptions

In 2004, Lawrence Berkeley National Laboratory (LBNL) published the results of a study, "Understanding the Cost of Power Interruptions to U.S. Electricity Consumers." (LBNL 2004)

The blackout in the northeastern United States and Canada on August 14-15, 2003 rekindled public interest in the reliability of the electricity grid. Power system sources suggested that investments of $50 to $100 billion would be needed to modernize the grid. LBNL attempted to quantify an important piece of information that has been missing from these discussions: how much do power interruptions and power quality events cost U.S. electricity consumers?

The framework relies on a simple mathematical expression that determines the economic cost of power interruptions (or power quality) as follows:

$$\text{Cost of power interruptions} = \sum_{i=1}^{m}\sum_{j=1}^{n} C_{Lj} \times E_{Lj} \times O_{Lj}$$

Where,

C = total number of electric power customers in each region and customer class sector

E = the frequency of power outage events in one year for each region and customer class sector

O = the cost per outage by customer class for each region

m = the number of customers in each customer class

n = the number of regions

i,j = indices for customer class and region, respectively

This expression can also be used to estimate the cost of power quality using costs per outage per customer that reflect these short-term disturbances. The simplicity of this formula belies the complexities involved in estimating the value of each of the three variables in the equation.

LBLN developed a comprehensive end-use framework for assessing the cost to U.S. electricity consumers of power interruptions and power quality events. This framework expresses annual power-interruption and power quality costs (referred to collectively as "reliability events") as a function of the:

- Number of customers by class and region;
- Duration and frequency of reliability events experienced annually (including both power interruptions and power quality events) by customers;
- Cost of reliability events, by event type, customer class, and region; and
- Vulnerability of customers to reliability events.*

*LBNL concluded that the vulnerability of customers to reliability events is included because it is an important component of the cost of reliability events. However, because there are no reliable, current data on customer investments in reliability-enhancing technologies (e.g., back-up generation, batteries, power-conditioning equipment), this component is not currently incorporated in their estimates of sensitivity analyses.

LBNL used the framework to review previous estimates of the national cost of power interruptions and power quality, including those developed by the Electric Power Research Institute (EPRI) and the U.S. Department of Energy (DOE), which range from $26 billion to $400 billion annually. LBNL's analysis shows that key assumptions underlying these early estimates reveal potentially significant biases; many of these biases cannot be fully understood until better information is collected than is currently available on the elements that contribute to the costs of reliability events.

Following LBNL's review of estimates prepared prior to 2004, they used the best information available at the time to develop a new estimate of the national cost of power interruptions. LBNL's work does not include power quality events. LBNL concluded that the cost of power interruptions is approximately $80 billion annually as shown in Figure 6-4 broken down by customer class.

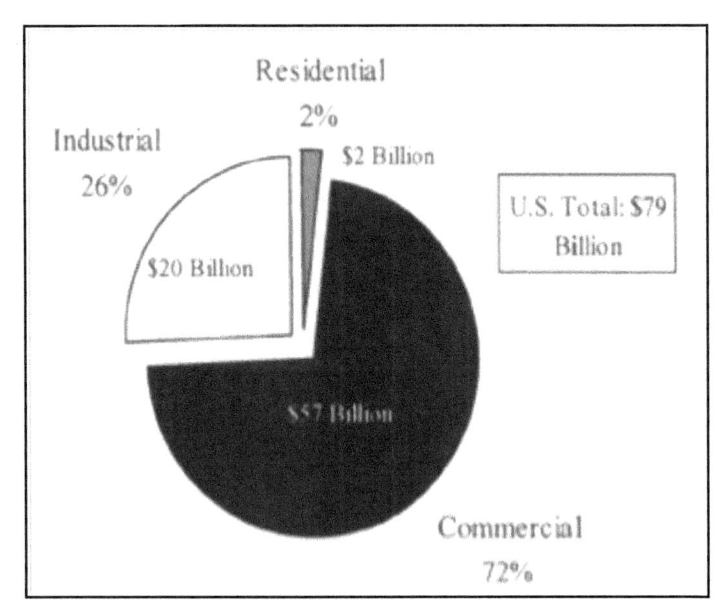

Figure 6-4. LBNL Base Case Estimate of the Cost of Power Interruptions by Customer Class (LBNL 2004)

LBNL's analysis shows that:

- The majority of outage costs are borne by the commercial and industrial sectors;
- As a result, although there are important variations in the composition of customers within each region, the total cost of reliability events by region tend to correlate roughly with the numbers of commercial and industrial customers in each region; and
- Costs tend to be driven by the frequency rather than the duration of reliability events.

Related to this last finding, their work revealed the importance of short-term, momentary interruptions, which last five minutes or less. LBNL concluded that (more frequent) momentary power interruptions have a stronger impact on total cost of interruptions than (less frequent) sustained interruptions, which last five minutes or more.

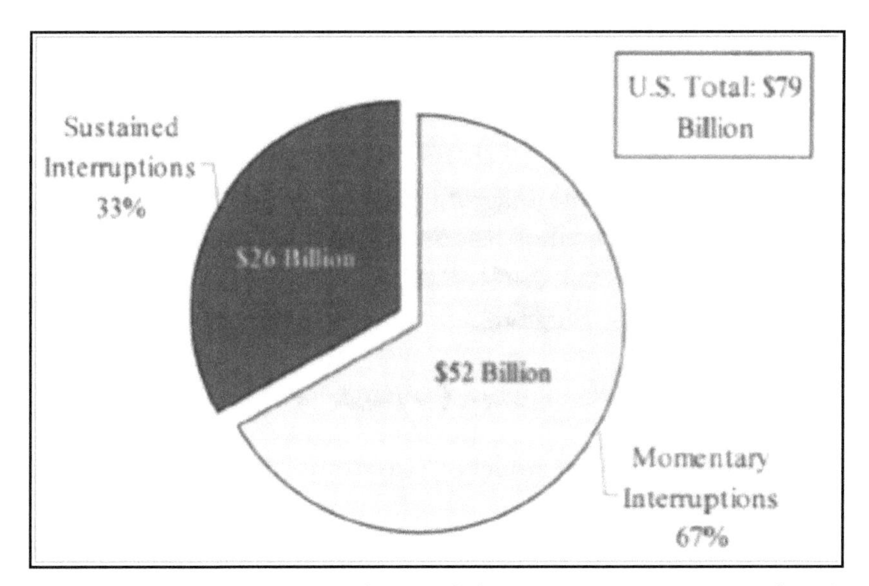

Figure 6-5. LBNL Base-Case Estimate of the Cost of Power Interruptions by Type of Interruption (LBNL 2004)

Value of Electricity: Based on LBNL Analysis
Hypothesis: Both momentary and sustained interruptions have a substantial impact on the Nation's economy
Assumptions: Momentary interruptions = $52 Billion/Year and Sustained interruptions = $26 Billion/year (2004)
Value of Electricity: Based on interruptions = $98 Billion/year (2014)

Consistent with LBNL's review of prior estimates, they found that there are significant gaps and uncertainties in the information currently available to support any estimate of the national cost of power interruptions.

LBNL performed a sensitivity analysis of their base case in which they varied key parameters used in the calculation in order to quantify the impact of these variations on the results. Figure 6-6 shows LBNL's resulting total cost of power interruptions for each of the following variations:

- Assuming that the duration and frequency of reliability events varies by region, based on the limited region-specific data collected;
- Assuming that the duration and frequency of reliability events is one standard deviation greater and less than the values used in the initial estimate, based on the total sample of data collected;
- Assuming that all outages are valued based on the assumption that they occur on a summer weekday afternoon or summer weekend night; and
- Assuming that the commercial and industrial sectors experience a disproportionately lower duration and frequency of reliability events than the resident sector.

Outage Example: The Impact of Hurricanes on Entergy's Service Area

During the period August 1992 to September 2005, Entergy*

---

*Entergy is an energy company engaged primarily in electric power generation and retail distribution operations.

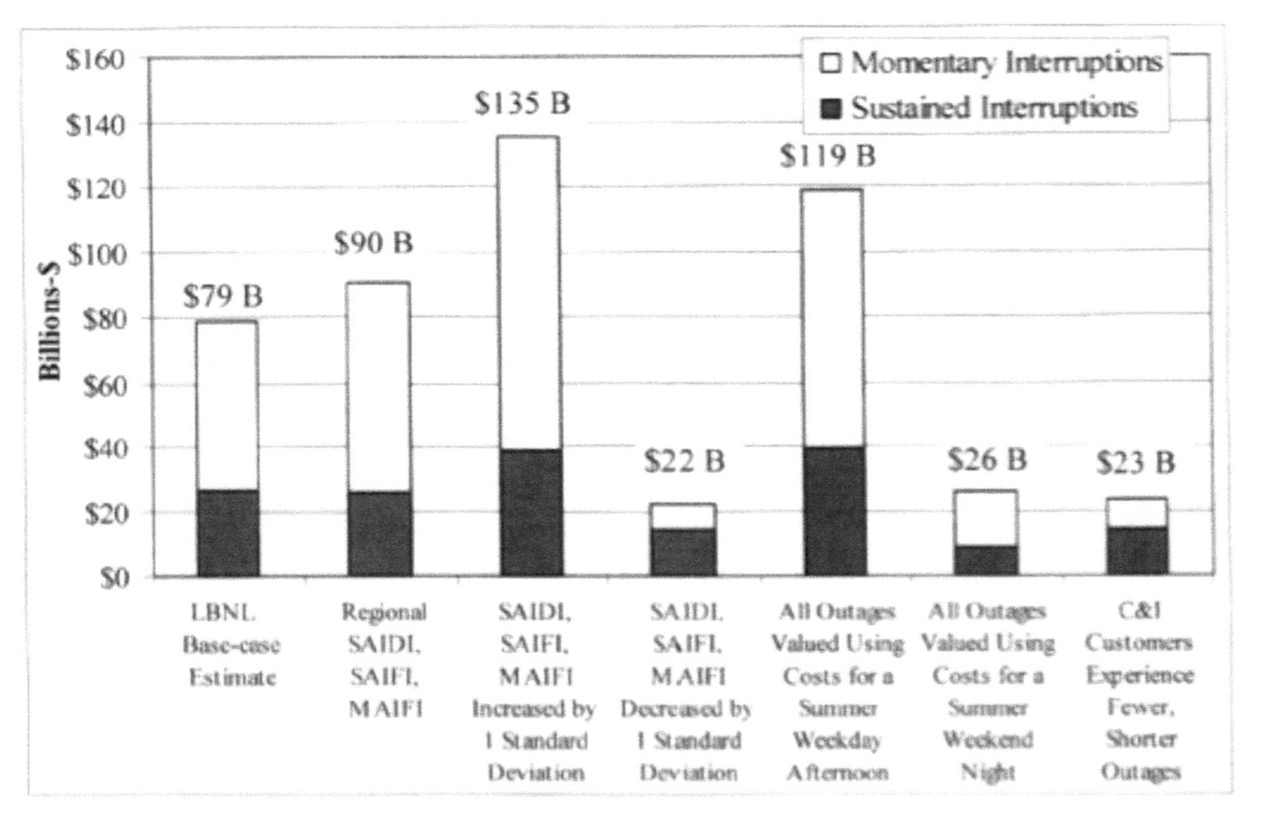

Figure 6-6. Summary of U.S. Cost of Power Interruption Sensitivity Cases (LBNL 2004)

experienced eight major storms, six of which were hurricanes. (Hintz 2006) These were as follows:

- Hurricane Andrew (Aug 1992) 250,000 customer outages
- Ice Storm (Feb 1994) 240,000 outages
- Hurricane Georges (Sep 1998) 260,000 outages
- Dual Arkansas Ice Storms (Dec 2000) 236,000; 247,000
- Hurricane Lili (Oct 2002) 243,000 outages
- Hurricane Cindy (Jul 2005) 289,000 outages
- Hurricane Katrina (August 2005) 1.1 million outages
- Hurricane Rita (Sept. 2005) 800,000 outages

The most dramatic example of outages comes from the double 2005 punch of Hurricane Katrina followed 25 days later by Hurricane Rita. In particular, Hurricane Katrina offered unique challenges since Entergy's own corporate headquarters had to be evacuated, and many employees' homes were destroyed. This was compounded by security threats in New Orleans, flooded gas facilities, contractors' bankruptcy fears and massive scale and logistical challenges.

Hurricane Rita offered a number of challenges as well. It was the second worst storm in the company's history—800,000 outages wherein massive damage was done to the transmission system and generation plants were damaged and isolated. In Texas, there were three days of rolling blackouts required for 142,000 customers in order to maintain the system.

Value of Electricity: Consumers without power during major storms

Hypothesis: In certain regions of the U.S. storms cause massive economic loss
Assumptions: Combined impact of hurricane Katrina and Rita cause loss of power to over 1 million customers; some for over 45 days
Value of Electricity: Loss of power to over I million customers for periods exceeding 45 days for some

Figure 6-7 illustrates the combined restoration profile of Katrina and Rita.

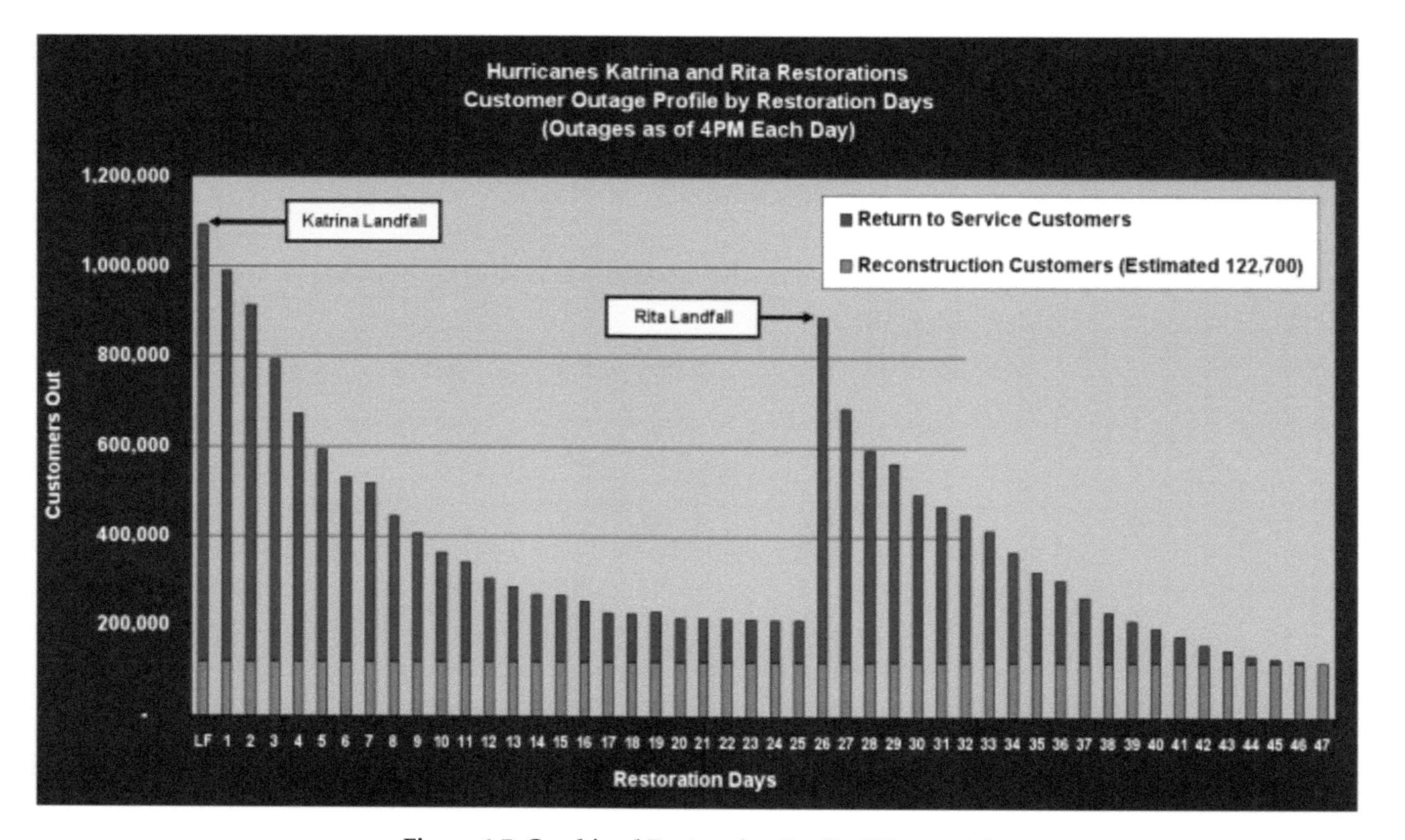

Figure 6-7. Combined Restoration Profile (Hintz 2006)

Superstorm Sandy

Superstorm Sandy hit the Atlantic seaboard in late 2012. Sandy is estimated to be the second costliest storm on record, behind only Hurricane Katrina. Over 8.5 million customers from Delaware to Massachusetts lost power. New Jersey and New York were hit particularly hard. In some regions, it took weeks or even months—especially in the coastal areas—to restore power, despite the efforts of 65,000 workers deployed from around the US and Canada. The long downtimes, while not unusual by historical standards, are now considered unacceptable by the public for a modern society and economy.

A few bright spots were evident amidst the catastrophe. At Princeton University, thanks to on-site generation, power went out only briefly before being restored. Thanks to a local microgrid, power remained on for the New York University campus in the middle of Manhattan, even as the grid went down around it. At Co-op City, a large housing development in the Bronx, 60,000 residents continued to have access to power throughout the storm. Again, thanks to on-site generation.

What kept the lights on in these locations, even as more robustly protected areas such as data centers, airports, and hospitals experienced power and fuel outages? These sites had installed *microgrids*—coordinated systems comprising distributed generation, storage, and controllable loads—to augment normal power from the grid. In all of these instances, microgrids had been designed to provide efficiency advantages through cogeneration of heat and power (CHP), but after Sandy, it became clear that load management with CHP also provided precious additional reliability and resiliency.

Superstorm Sandy made landfall near Atlantic City, New Jersey, on October 29, 2012, and then continued northwest over New Jersey, Delaware and Pennsylvania. The heaviest damage was due to record floods. A storm surge of 12.65 feet hit New York City. New Jersey experienced a storm surge of 8.57 feet. In all, the storm damaged 650,000 homes and knocked out power for 8.5 million consumers. (Executive Office of the President 2013)

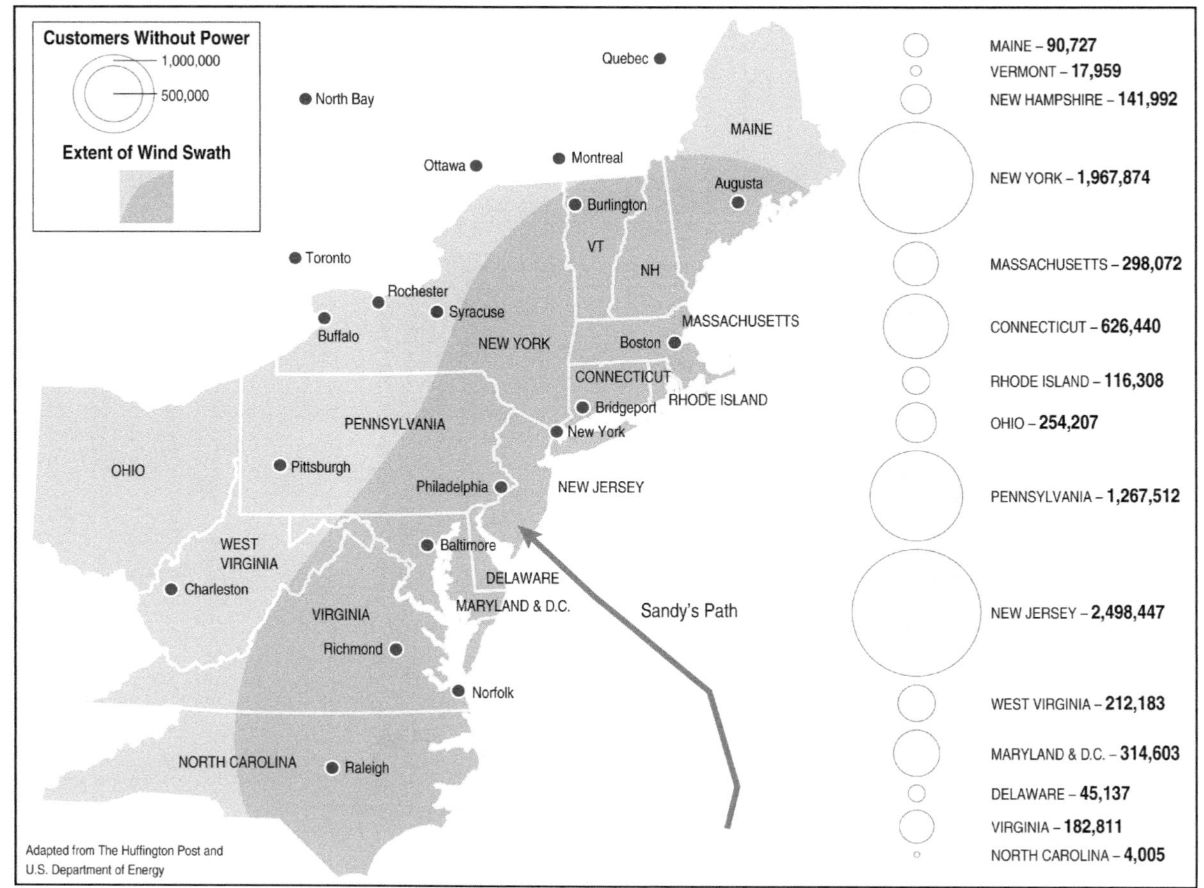

Figure 6-8. Hurricane Sandy Power Outage Map (Huffington Post 2012)

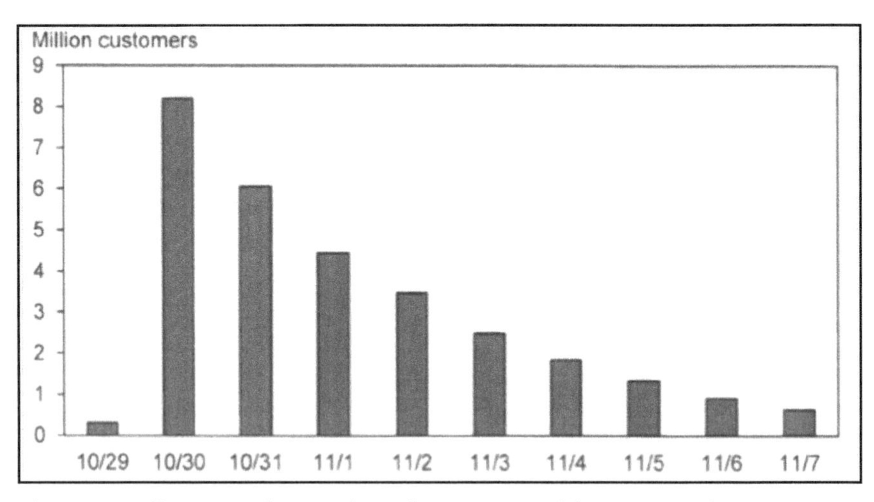

Figure 6-9. Illustrates the number of customers without power in New Jersey and New York as a result of Superstorm Sandy (Executive Office of the President 2013)

Value of Electricity: Based on Superstorm Sandy
Hypothesis: Superstorm Sandy caused massive power outages
Assumptions: Massive damage to infrastructures from hurricane Sandy
Value of Electricity: 8.5 million consumers without power in New York and New Jersey

Following Hurricane Sandy, power outages impacted approximately 8.5 million customers, affecting millions more people. Additionally, breaks in natural gas lines caused fires resulting in the destruction of many residences. Access to gasoline and diesel fuel was severely impaired following Sandy. This was largely caused by flooding damage to major terminals and docks in New Jersey. These fuel shortages delayed first responders. As a result, portable generators sat unused and lines at fueling stations were long, while consumers struggled to identify which gas stations had power and were operational. (USDH&UD 2013)

THE COST OF OTHER POWER DISTURBANCES

It is widely known that power system limitations are manifesting in reliability problems and costly disturbances. A number of examples include the following expressed in 2014 dollars (EPRI 1006274):

- 20-minute outage at one Hewlett-Packard fabrication plant — $38 million loss
- 20 minute outage at Sun Microsystems — $25 million loss
- Momentary voltage sags at baby food factory — $50,000 loss

Total U.S. power reliability losses per year — $150-237 billion

Value of Electricity: Based on surveys of commercial and industrial consumers
Hypothesis: Power outages and power quality adversely impact the U.S. economy
Assumption: Consumers were asked about their actual losses
Value of Electricity: Total U.S. losses range between $150 and $237 billion per year

Figure 6-10 and 6-11 illustrate the results of the most thorough surveys ever conducted on this impacts of poor power quality and reliability.

Canada-U.S. Blackout, August 2003

The massive blackout in the northeastern United States and Eastern Canada on August 14-15, 2003 rekindled public interest in the reliability of U.S. electricity service. Following the blackout, the U.S. electricity system was called "antiquated" and likened to that of a third-world nation, and industry sources suggested that investments of $50 to $100 billion would be needed to modernize the grid. Experts called for the development of a framework for systematically and rationally assessing the economic costs of power interruptions and power quality on U.S. electricity consumers.

During the past decade, there have been several efforts to assess the economic cost of power interruptions and power

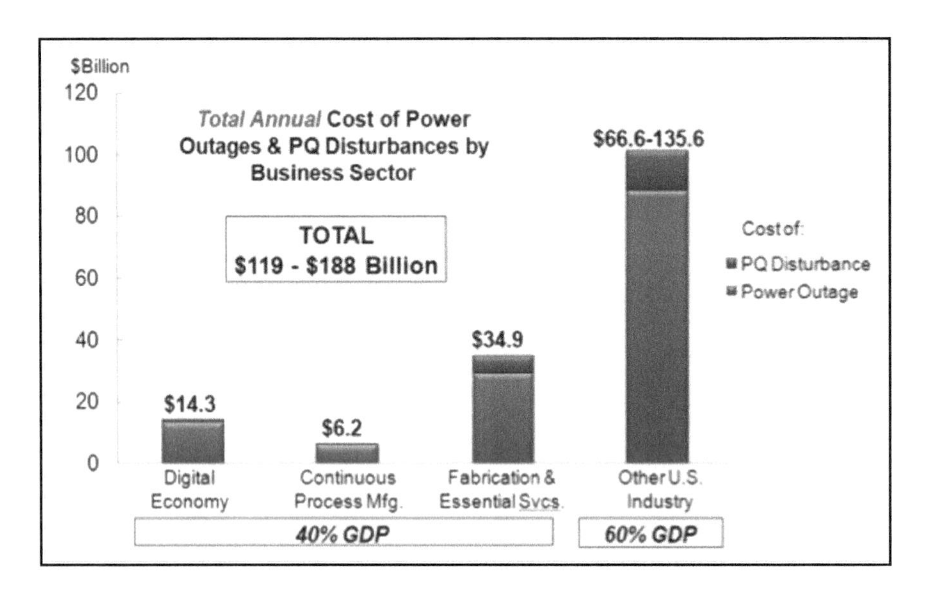

Figure 6-10. A Toll Felt Throughout the U.S. Economy (Primen Study: The Cost of Power Disturbances to Industrial & Digital Economy Companies, EPRI 1006274)

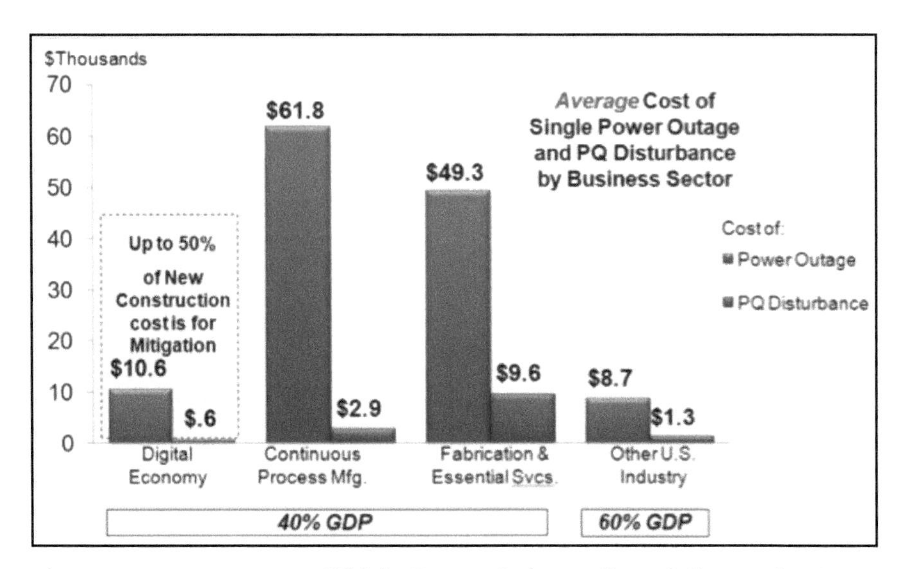

Figure 6-11. Impacts Vary Widely Due to Industry Type & Preventive Measures (Primen Study: The Cost of Power Disturbances to Industrial & Digital Economy Companies, EPRI 1006274)

quality. During the 1990s, the Electric Power Research Institute (EPRI) estimated the national cost of $26 billion per year based on a figure that had been presented at a power quality conference. Later, EPRI extrapolated from this figure and began estimating the power interruption cost at $50 billion per year. During the same period, a U.S. Department of Energy (DOE) study offered estimates ranging from $150 to $400 billion per year, based on an extrapolation from a utility value-of-service study. Finally and most recently, EPRI prepared a new set of estimates ranging from $119 billion to $181 billion per year; $119 billion per year is the figure most often quoted from that study. (Primen 2001)

In a subsequent study, the Electricity Consumers Resource Council (ELCON) compiled reports on the impacts of the August 2003 blackout. (ELCON 2004) The information was based on trade press or media coverage of the blackout. Table 6-1 summarizes the impacts compiled by ELCON.

Value of Electricity: Value of electricity to major industries based on Superstorm Sandy
Hypothesis: Major industries suffer substantial loss in major storms
Assumptions: ELCON survey of members
Value of Electricity: $142.4 Million/day

Southwest Power Outage

A widespread power outage in 2011 which affected southern California as well as parts of Arizona and parts of the country of Mexico was estimated to have brought $100 million in economic losses. This estimate is based on an analysis by the National University System Institute for Policy Research that businesses lost about $70 million, largely because they had to close. In addition, overtime for government workers cost up to $20 million and the cost of food spoilage was estimated to be $18 million. (Jergler 2011)

Hurricane Irene

Hurricane Irene made landfall near Cape Lookout, North Carolina, on August 27, 2011 and then continued north-eastward

Table 6-1. Compiled Impacts of the 2003 Blackout (ELCON)

Corporation	Industry Type	Plants Impacted	Workers Idled	Production Impacted	Total Impact ($)
General Motors	Automotive	19	100,000		–
Ford Motor Co.	"	23			–
Daimler Chrysler	"	14		10,000 vehicles scrapped	50 M
Honda	"	1			–
Various	"	4			–
Manhattan Oil	Petroleum	1		76,000 barrels/day (bpd)	7.6 M/day
BP Plc.	"	1		160,000 bpd	16 M/day
SUNOCO	"	1		140,000 bpd	14 M/day
Imperial Oil	"	2		237,000 bpd	23.7 M/day
Petro-Canada	"	1		90,000 bpd	9 M/day
Shell Canada	"	1		75,000 bpd	7.5 M/ day
Suncor Energy				70,000 bpd	7M/day
Marathon Oil	"	1		76,000 bpd	7.6 M/day
Various	Steel	11			–
Nova Chemicals	Chemical	1		150M lbs. chemicals	–
Various	"	11			–
Total					142.4 M/day

making a second landfall near Atlantic City, New Jersey, affecting more than 6.5 people in the United States. Irene caused the death of 41 people in the U.S. and resulted in $15.8 billion in total damages (Avila and Cangialosi 2011)—the seventh costliest hurricane in U.S. history.

A recent review of cost of outage studies conducted by EPRI (EPRI 106082) estimated costs to consumers for one hour of interrupted service on a summer weekday afternoon. The researchers found that the average cost for small commercial and industrial customers was $373 per unserved kilowatt hour (kWh). For large commercial and industrial customers, the average cost was $25 per unserved kWh, and for residential customers, the average cost was $2.60 per unserved kWh.

Value of Electricity: Based on Galvin Electricity Initiative Estimates
Hypothesis: Consumers spend at least 50 cents per day on other goods and services to cover the cost of outages
Assumptions: Electricity from grid- related services is 99.7% reliable
Value of Electricity: $150 Billion/year

Galvin Electricity Initiative Estimates

According to the Galvin Electricity Initiative, the U.S. electric power system is designed and operated to meet a "3 nines" reliability standard (99.9). According to Galvin, electric grid power is 99.97% reliable. Galvin alleges that in practice, it translates to interruptions in the electricity supply that cost American consumers an estimated $150 billion a year. Galvin estimates that for every dollar spent on electricity, consumers are spending at least 50 cents on other goods and services to cover the costs of power failures.

These costs result from losses in affected industries being passed down to consumers. For example, broken down by business type, the average estimated cost of a one-hour power interruption is detailed in Table 6-2.

Value of Electricity: Based on Southwest power outage in 2011 and Hurricane Irene
Hypothesis: Power outages cause substantial economic harm
Assumptions: Southwest power outage cost $100 million and hurricane Irene $15.8 billion
Value of Electricity: One outage can cost nearly $16 billion

Table 6-2. Galvin estimates of the average cost of a one hour interruption

Industry	Average Cost of 1-Hour Interruption
Cellular communications	$41,000
Telephone ticket sales	$72,000
Airline reservation system	$90,000
Semiconductor manufacturer	$2,000,000
Credit card operation	$2,580,000
Brokerage operation	$6,480,000

Galvin believes that in an increasingly digital world, even the smallest power quality event or outages can cause loss of information, processes and productivity. Interruptions and disturbances measuring less than one cycle (less than $1/60^{th}$ of a second) are enough to crash servers, computers, intensive care and life support machines, automated equipment and other microprocessor-based devices. Galvin suggests that problems of power outages and their associated costs are becoming more severe. Forty-one percent more outages affected 50,000 or more consumers in the second half of the 1990s than in the first half of the decade. The "average" outage affected 15% more consumers from 1996 to 2000 than from 1991 to 1995 (409,854 versus 355,204).

Data Center Outages

A 2013 study on data center outages sponsored by Emerson Network Power (Ponemon 2013) highlighted concerns regarding data center outages. The study concluded that unplanned data center outages present a difficult and costly challenge for organizations as they directly impact the availability of data centers. The study followed one conducted in 2010 where findings revealed that organizations were underestimating the impact unplanned outages have on their operations. The 2013 study revealed that this is changing and in some respects show that the ability to prevent data center outages is improving. New innovations around monitoring and Data Center Infrastructure Management (DCIM) are expected to have a positive impact on data center availability for the industry.

Ponemon surveyed 584 individuals in U.S. organizations who have responsibility for data center operations. Eighty-five percent of respondents report their organizations experienced outages in the past 24 months. Among those organizations that had a loss of primary utility power, 91% report their organizations had experienced an outage.

The findings suggest companies do not have practices to reduce or respond to outages. Seventy-one percent of respondents agree their company's business model is dependent upon data centers to generate revenue and conduct ecommerce. Most organizations in the study have had at least one unplanned outage in the past 24 months. Respondents averaged two complete data center outages over the two-year period, with an average duration of 91 minutes. The duration of the data center outages correlates to lack of resources and planning as only 38% agree there are ample resources to bring their data center up and running if there is an unplanned outage.

Companies in e-commerce and IT services reported the lowest number of complete data center outages (1.63 and 1.69, respectively), while health care had the highest number with 2.7 total data center outages every two years. Public sector organizations reported the longest durations, and e-commerce has the shortest duration of outages.

Figures 6-12 compares the extrapolated duration of complete data center outages by industry.

In more recent years, the potential impact of blackouts on data centers has grown phenomenally. Not merely because of the growth of the importance of the internet and social networking, but more recently due to the evolution of the cloud as homes and businesses are moving from dependency on PCs to the use of laptops and now to the adoption of the smart phone and of tablets.

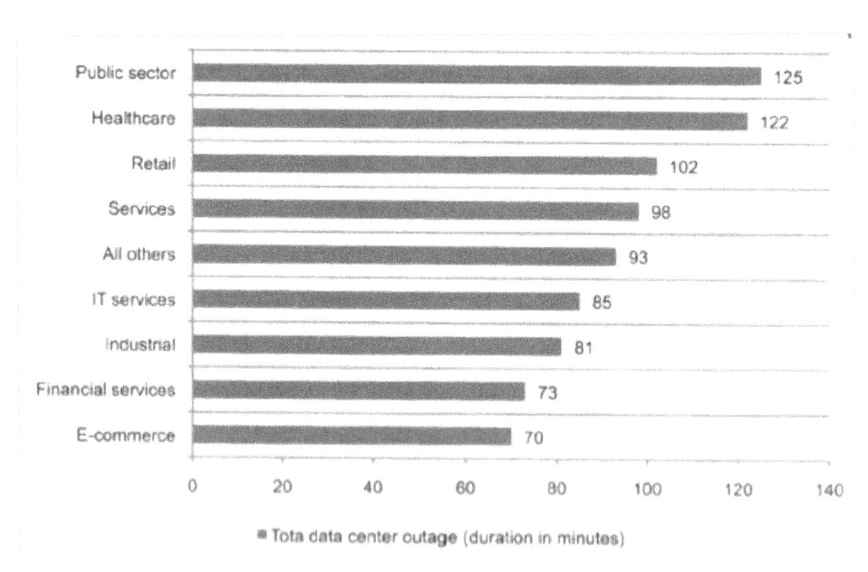

Figure 6-12. Extrapolated Duration of Complete Data Center Outages by Industry Segment (Ponemon 2013)

As society moves to these smaller and more portable devices, use of the cloud will explode in order to enable the expanding use of "apps."

Sears Blackout Cost Millions of Dollars

Sears Holdings Corporation experienced two power failures early in 2013 at its main data center in Troy, Michigan, claiming it cost more than $2 million in lost profits and another $2.8 million to fix, according to a lawsuit filed May 24th in Cook County Circuit Court. The problems began January 3rd with the failure of one of four uninterruptable power supplies. All four power supplies subsequently failed, followed by the failure of a bypass power setup, shutting down Sears' computer systems, including its web site, before it could restore power and bring the computers back up.

The five-hour failure cost Sears $1.58 million in profit, according to the lawsuit. The server farm ran on generators for eight days, burning through $189,000 in fuel. Three of the four power supplies failed again on January 24th. (Pletz 2013)

Value of Electricity: Data center outages
Hypothesis: Data centers are among the fastest growing and most electricity dependent consumers Assumptions: Data centers experience between 65 and 125 minutes interruption in primary electricity supply each year
Value of Electricity: A Sears data center five hour power failure cost over $1.7 million

Lessons from Africa

A more recent study examined the value of electricity as observed from outages experienced among African firms. (Oseni) The study's approach examined the economic costs of unsupplied electricity using evidence from backup generation. The study observed that electricity provision in Africa has been marred by lack of generation expansion, poor fuel supply and frequent power outages. One of the strategies often adopted by firms to cope with this poor supply is investment in backup generation. African outage cost valuation studies have so far ignored the effects of firm's characteristics on the extent to which a firm can suffer from power outages. Using cross-sectional data of 7353 firms currently operating in 12 African countries, researchers investigated the extent to which firms' characteristics affect the costs of power outages and how that might create incentives to invest in backup generation. They used three different methods including marginal cost, incomplete backup and subjective evaluation. The results revealed that while the demand for backup is influenced by power outages, firm's size, manager's experience, other factors such as firm's reputation and the use of the internet for firm's operations play more roles in decision to invest in backup.

The results also revealed that unmitigated costs still account for the larger proportion of the total outage costs despite high prevalence of backup ownership among the firms. The results also indicated that the previous studies on Africa underestimated the costs of power outages due to the omission of additional costs that may have resulted from incomplete investments in backup. The results revealed that while marginal cost method underestimates outage cost, subjective evaluation had a potential to overstate it. An average firm in Africa incurs outage cost ranging between

$0.46 and $1.25 per kWh of unsupplied electricity accumulating to between $1,343 and $3,650 per kW. The analysis further suggests that firms can still benefit significantly even when the current subsidized tariffs are replaced by cost-reflective rates that guarantee stable electricity supply.

Value of Electricity: Based on African studies
Hypothesis: Cost of unserved energy to African firms
Assumptions: Loss of power causes economic losses
Value of Electricity: Africa outage costs range between$0.46 and $1.25/kwh

VALUE OF LOST LOAD

Value of lost load (VOLL) is a consumer facing metric which measures one element of the performance of a power system. It is used as a planning tool to estimate the economic damage ($ in the U.S.). It typically estimates or elicits damage or inconvenience information from consumers and is often expressed in dollars per kilowatt-hour. The total damage associated with any outage is the product of "loads" or consumer demands for electricity which remain unmet or unserved during an outage times the VOLL.

A basic estimate of VOLL can be found by dividing gross domestic product by total energy consumed. US Gross Domestic Product—the market value of goods and services is $17,090 billion. (U.S. Department of Commerce, Bureau of Economic Analysis—www.bea.gov March 27, 2014). According to the US Department of Energy Information Administration "Electricity Retail Sales to Ultimate Customers in 2012 was 3.694 x 109 kWh. (www.eia.gov). Assuming that very little, if any of the GDP would have been produced if electricity was not available, one could infer the VOLL for the USA would be $17.09 x 1012/3.694 x 109 kWh = $4,626/kWh.

Value of Electricity: Electricity drives GDP
Hypothesis: All GDP is dependent on electricity
Assumptions: No electricity was produced or was available
Value of Electricity: $4,626/kwh (2012)

There is a relationship between the cost of electricity and the quality of service. This can be used to estimate or infer the value of electricity. In the electricity business, there is some form of political oversight or regulation. This theoretically encourages suppliers to maintain quality to some optimum level (OQL). Some assume that this can be achieved by measuring system reliability indices and then reviewing the effects of various incentive and penalty mechanism which might stimulate improvements to achieve OQL.

LBNL VOLL Study

In a Value of Lost Load study, LBNL estimated the lengthiest outage durations as $0.92 per unserved kWh for residential customers and $11.29 per unserved kWh for medium and large commercial and industrial customers. These represent the values to customers of avoiding the loss of power—that is, estimates of the economic damages that they would incur as a result of a power outage lasting 8 hours. Of course, these values are not the cost basis to the electric utility, nor are they the price than an electric utility would charge its customers. (LBNL 2004)

Value of Electricity: Based on LBNL Study of VOLL
Hypothesis: Value of Lost Load
Assumptions: Estimates based on LBNL surveys of other studies
Value of Electricity: $.092/kwh for residential customers and $11.29 for medium and large commercial and industrial customers

Lessons from Europe

A 1993 British study estimated sector costs expressed as damage functions (Cranton 2000). Table 6-3 details these findings (here expressed in dollars/kWh of peak demand) (Cranton 2000).

Value of Electricity: Customer damage function
Hypothesis: Customers experience damages from loss of power
Assumptions: Based on 1993 British study
Value of Electricity: Ranges from $2.40 to $365.50/kwh for outages from <1 second to 24 hours

Table 6-3. 1993 British Study—Sector Customers Damage Functions. (1999) dollars/kWh of Peak Demand

Table – 1993 British Study
Sector Customers Damage Functions
(2014) dollars/kwh of Peak Demand

Customer Category	Outage Duration						
	<1 sec	1 min	20 min	1 hrs.	4 hrs.	8 hrs.	24 hrs.
Industrial	15.0	15.7	34.8	61.4	175.6	292.0	365.5
Commercial	2.4	2.4	9.4	25.9	95.0	191.2	243.1
Residential							
Large Users	16.5	16.5	16.7	17.5	21.6	23.6	32.5

Tables 6-3 and 6-4 depict the results of a study by UMIST (Strbac, 1999 estimated Customer Interruption Costs in U.K).

Table 6-4. UMIST Estimates of Interruption Costs ($/kWh) [2014]

	1 min in one day	1440 min in one day
Residential	0.19	.25
Commercial	1.86	92.80
Industrial	11.60	116.00
Large User	2.32	4.64

Value of Electricity: Based on UK study estimating customer interruption costs
Hypothesis: Consumers can estimate the damage they would incur from interruptions
Assumptions: Customer survey
Value of Electricity: Ranges from $.19 to $116.00 per kWh

Finland

A study in Finland in 1997 asked consumers to estimate the costs to themselves of supply interruptions of varying characteristics.

In 1995 prices, it estimated £3.7/kWh or $8.88/kWh (2014) for a one hour long outage for industrial users falling to £1.3/kWh or $3.12/kWh (2014) for a 24 hour outage.

Value of Electricity: Based on Finish study
Hypothesis: Survey can be used to estimate customer losses from an outage
Assumptions: Survey results
Value of Electricity: Ranges from $3.12 to $8.88/kwh

The ICE Calculator

The U.S. Department of Energy (DOE) has developed a tool to assist planners to estimate damages from outages. These so-called "damage functions" vary by time of day, season of the year and duration of outage. The basis for the tools is extensive surveys conducted by utilities over decades. These data reveal the Value of Lost Load (VLL), usually $5,000-$15,000/kWh. (Brown, 2013)

Table 6-5 depicts the Estimated Average Electric Customer Interruption Costs by Customer Type and Duration (Summer Weekday Afternoon). Source: Sullivan

Value of Electricity: Based on DOE calculator
Hypothesis: A meta-analysis of studies by others will allow a method to estimate interruption costs
Assumptions: Losses differentiated by duration and type of consumers can provide reasonable estimates
Value of Electricity: Range between $.99 and $2665/kwh

London Economics Study

London Economics summarizes a series of studies which would help categorize people's opinions about value. When asked, "In the event of a four-hour outage at the worst time of next year, what actions would you take?" Choices and associated costs included:

Table 6-5. DOE ICE Calculator Results

Interruption Cost ($)	Momentary	30 mins.	1 hour	4 hours	8 hours
Medium and Lg C&I					
Per event	13,049	17,437	22,560	65,699	104,218
Per un-served kWh	192.14	42.74	27.75	20.20	15.98
Small C&I					
Per event	484	677	908	2993	5293
Per un-served kWh	2665	618	414	341	301
Residential					
Per event	3.00	3.66	4.33	8.66	11.88
Per unserved kWh	26.61	4.84	2.89	1.44	0.99

- Light candles or use torch (flashlight) for four hours ($4.00)
- Buy gas lantern ($10.00)
- Buy some ice to put into refrigerator ($2.00)
- Drive to a relative or friend's home and stay with them ($10.00)
- Buy portable gas stove for cooking and boiling ($30.00)
- Buy back-up battery supply to use computer for up to 30 minutes ($40.00)

- Buy portable kerosene or LPG space heater to provide heating for one room ($50.00)
- Go to a restaurant for one meal ($50.00)
- Do nothing and wait for power to return ($0)

Value of Electricity: Based on London Economics study
Hypothesis: Residential consumers asked what actions they would take in an anticipated 4 hour outage
Assumptions: Residential consumers take actions to protect themselves or occupy their time
Value of Electricity: Ranged from $0 impact to $50.00 per residential customer

Another study which refers to the value of electricity was called the Value of Lost Load performed by London Economics for ERCOT and the Texas Public Utility Commission dated June 17, 2013. Their literature review found that average VOLLs for a developed, industrial economy range from approximately $9,000/MWh to $45,000/MWh. On a more disaggregated level, residential customers generally have a lower VOLL ($0/MWh to $17,976/MWh) than commercial and industrial (C/I) customers (whose VOLL range from about $3,000/MWh to $53,907/MWh).

Value of Electricity: Based on ERCOT study of VOLL
Hypothesis: Developed, industrial economy
Assumptions: Range of values depending on customer type and length of outage
Value of Electricity: Residential $0/kwh to $17.98/kwh; commercial and industrial ranges from $30.00/kwh to $539.07/kwh

Value of Electricity: Based on London Economics study of a variety of countries
Hypothesis: Value of loss load for various countries and one U.S. TSO
Assumptions: Survey of various countries
Value of Electricity: $107/kWh for residential; $82.39 to $170.13/kWh for commercial and industrial depending on size

Table 6-6 summarizes the London Economics study by region. Table 6-7 summarizes the compilation of studies reviewed by London Economics.

Table 6-6. Summary of VOLLs by Jurisdiction (London Economics Study)

Report/Market	Methodology	System-wide VOLL	Residential	Non-Residential		Applicability to ERCOT
				Large C/I	Small C/I	
U.S.-Southwest	Analysis of past survey results		$0	$8,774	$35,417	High
U.S.-MISO	Analysis of past survey results Macroeconomic analysis		$1,735	$29,299	$42,256	Moderate
Austria	Survey		$1,544			Low
New Zealand	Survey	$41,269	$11,341	$77,687	$30,874	Low
Australia-Victoria	Survey	$44,438	$4,142	$28,622	$10,457	Moderate
Australia	Analysis of past survey results	$45,708				Low
Republic of Ireland (2010)	Macroeconomic analysis	$9,538	$17,976	$10,272	$3,302	Low
Republic of Ireland (2007)	Macroeconomic analysis	$16,265				Low
U.S.-Northeast	Macroeconomic analysis	$9,283-$13,925				Low

*All values in 2012 U.S. dollars/MWh

Table 6-7. Summary of compilation of studies reviewed by London Economics

Study	Commercial & Industrial ($/kwh)	Residential ($/kwh)
MISO	292.99 – 422.56	1.74
Austria	41.15 – 272.60	41.15 – 272.60
Austria (12 hour outage)	50.15 – 60.44	50.15 – 60.44
New Zealand	539.07	67.79
Australia	308.74 – 776.87	113.41

Chapter 7

Consumer Willingness to Pay

WILLINGNESS TO PAY

There are actually several methods that have been used to attempt to evaluate the benefits of electrification* and, in particular, rural electrification that have been applied to developing countries. One involves the concept of Willingness to Pay (WTP) which establishes a conservative benchmark, albeit likely much less than the actual value of electricity. In this method, citizens are asked to indicate what they would be willing to pay for electricity—if it was available or for improvements in electricity service.

The concept of Willingness to Pay (also known as W2P) is a term used by economists to characterize as person's willingness to pay for something representing the value they attach to it. Therefore, an individual's willingness to pay for one more unit of a good is a dollar measure of the benefits which the extra unit of goods gives them. WTP studies typically involve the use of surveys where consumers are asked their willingness to pay more for improved goods or services. Their response depends on the motivation of the respondent to provide information. Of course the quantity of complexity of the information provided can lead to respondents to disregard the information provided.

US CUSTOMERS WTP

In 2004 EPRI conducted an analysis of the business community's attitude toward additional investments in enhancing

*Electrification is the concept of providing electricity or electrifying a process, a building, a community or a country.

the power delivery system so as to mitigate disturbances. (EPRI 1011363) The project team surveyed a statistically representative sample of 401 U.S. businesses and conducted 25 additional qualitative, in-depth interviews with a combination of high profile energy users and businesses that had previously indicated a reluctance to fund such investments. The goal of the qualitative interviews was to get a better understanding of the factors that might make businesses less likely to support investments in the U.S. power system.

Figure 7-1 shows the percent of US business customers who would be willing to support an investment in enhancing the power delivery system, as a percentage of how that investment would affect their electric bills.

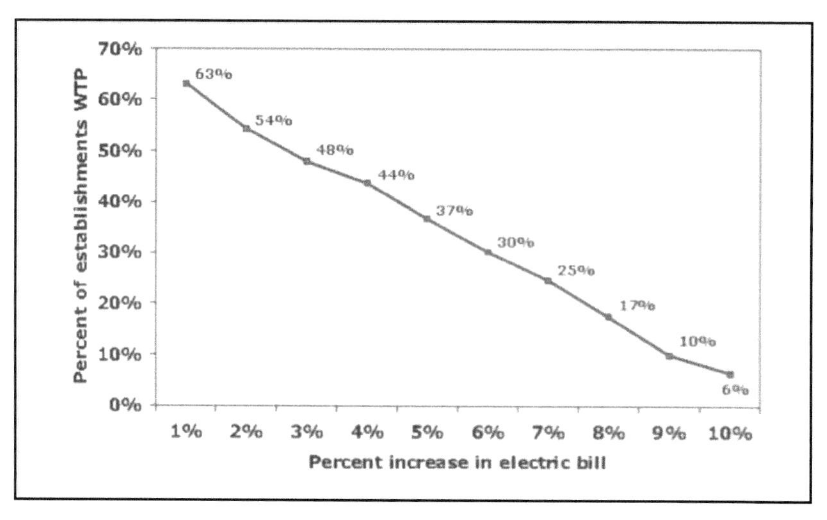

Figure 7-1. Percentage of Business WTP Various Increases on their electric bill to fund power systems enhancements (EPRI 1011363)

The survey results show that a majority of U.S. businesses (54%) would be willing to pay 1% or 2% more on their electric bills to fund an enhanced power delivery system that would provide improved power reliability and power quality, as well as enhanced capabilities for businesses to monitor and control their energy use. Roughly a third of businesses would be willing to pay up to

5% more to fund such investments. Only 6%, however, would be willing to pay 10% more to fund this investment—a funding level closer to the amount EPRI estimated would be needed to create a sustainable electric infrastructure for the 21st century. Customers willing to pay the most were found more often among commercial than manufacturing firms. They identify power reliability as their most important energy concern, although they have experienced fewer outages and power quality problems than other consumers. Indeed, consumers with the worst reliability and PQ experiences tend to be unwilling to pay more money for enhancements.

Figure 7-2 shows that roughly 65% of commercial customers (large and small) are WTP 1%. Only 59% of large manufacturing establishments, and 49.1 of small manufacturing would be WTP 1%. As the cost of the investment increases, WTP drops off for all four groups along a similar trajectory.

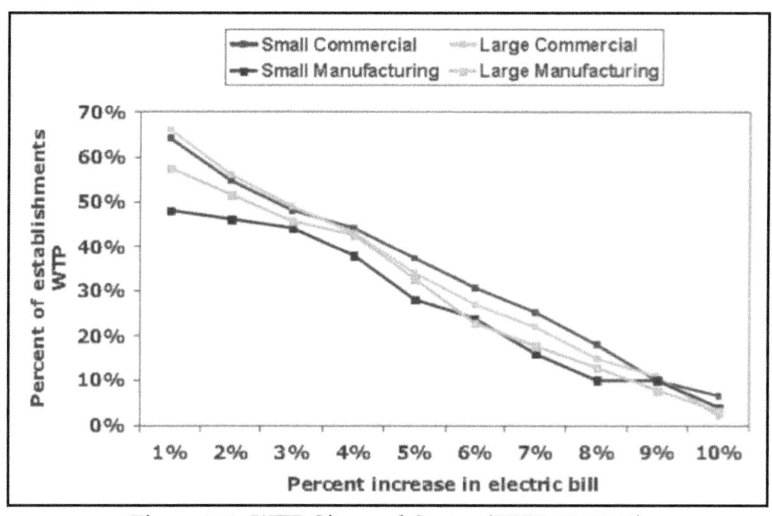

Figure 7-2. WTP Size and Sector (EPRI 1011363)

Among the reasons business customers are unwilling to pay for power system enhancements:

- Many of the benefits of an enhanced power system such as new energy management capabilities and better information about energy use are too vague and ill defined to be compelling. As a result, willingness-to-pay tends to rise and fall on the questions of reliability and power quality alone.

- Businesses have difficulty quantifying the cost of inadequate reliability and power quality
- Many customers have already made significant investments in standby generation and power quality solutions on their side of the meter, diminishing the benefits they would see from additional, utility-side investments.
- Many customers do not feel that they require greater reliability or power quality than the current system supplies.
- Some customers believe that utilities should be able to fund additional investments without raising rates.
- Some customers are reluctant to trust utilities with additional funds to make improvements often believing that such funds would be not be used for system improvements.

UK Households WTP

The University of Cambridge (Cambridge 2008) conducted studies of UK households' WTP a premium for fewer occurrence of blackouts. The studies in 2006 and in 2008. By using a survey approach to it, they directly asked people's willingness to pay for a "good" or attribute of a purchase. In this case for improvement in electricity reliability. In economists terms the question was whether they would be willing to pay monthly in order to secure a quality improvement? Two facets were explored:

1. An open ended contingent valuation: "How much will you be willing to pay per month to secure a quality improvement?

2. A closed ended contingent valuation: "would you be willing to pay $X to secure a quality improvement? (Yes or No)."

In these studies only a small percentage of households stated a WTP for avoidance of blackouts.

TVA Customers Willingness to Pay

A number of customer surveys are intended to either understand the experience they may have had during actual outages or to ask their willingness to pay to avoid an outage. The Tennessee Valley Authority (TVA) conducted a survey which estimated the premiums their customers would have paid to avoid an outage. The results were as follows:

$.04/kWh Residential
$2.46/kWh Industrial
$6.72/kWh Commercial

Value of Electricity: Based on TVA survey
Hypothesis: Customers could estimate their willingness to pay to avoid an outage
Assumptions: Customers will pay to avoid outages
Value of Electricity: Ranges from $.04/kWh residential to $2.46/kWh industrial and $6.72/kWh commercial

Asian Development Bank WTP

In 2002, the Asian Development Bank published a brief on Measuring Willingness to Pay for Electricity. In their brief they estimated a WTP among developing nations equivalent to the cost of kerosene based on equal energy content, which was $20/kWh. (Choynowski 2002)

Value of Electricity: WTP in developing countries
Hypothesis: WTP can be calculated based on the value of kerosene
Assumptions: Kerosene valued at an equivalent $20/kWh in 2002 or $26.40/kWh
Value of Electricity: $26.40/kWh

Sweden

Gutenberg University conducted an analysis based on a survey among Swedish households to determine the WTP in 2004.

The researchers observed that empirical studies typically treat electricity as one good of many consumed by households. Strictly speaking this is correct—households do consume elec-

Table 7-1. Predicted Swedish WTP in SEK/outage at various outage durations, 95% confidence intervals in parentheses. (Carlsson 2004)

Duration (Hours)	Planed outage	Unplanned Outage
1	6.30	9.39
	(4.30 - 10.32)	(7.52 - 15.95)
2	12.37	16.27
	(9.55 - 16.72)	(13.60 - 22.55)
3	18.18	25.19
	(15.74 - 25.95)	(21.97 - 32.56)
4	28.46	37.32
	(23.46 - 38.20)	(33.87 - 48.21)
5	38.45	48.76
	(32.41 - 49.44)	(45.27 - 64.43)
6	50.83	63.86
	(44.08 - 63.71)	(60.48 - 84.49)
7	66.01	83.36
	(57.89 - 81.67	(78.87 - 114.31)
8	84.42	108.09
	(75.01 - 103.95)	(100.69 - 153.87)

tricity. But the demand for electricity is a derived demand and is essentially an input to the production of services from a stock of electricity—consuming equipment in the household. Therefore, it can be argued that electricity is not part of a household's utility function. Rather it should be expected that it enters indirectly through the user cost with the services produced by the electricity consuming equipment.

This is based on a study of Willingness to Pay to avoid Power Outages (Carlsson 2004). In this study these households in populated areas experienced an average of 0.08 planned and 0.39 unplanned power outages per year with an average duration of 12 and 23 minutes, respectively. In sparsely populated areas they experienced .60 planned and 1.54 unplanned with corresponding durations of 83 and 203 minutes.

The results, as shown in Table 7-1 indicate a WTP of between 6.3 and 84.42 SEK for planned outages of 108 hours and 9.39 and 108.09 SEK for unplanned outages (2004 SEK). This translates to a WTP in 2014 dollars of between \$1.17 and \$14.96/kWh for planned outages and between \$1.26 and \$19.16/kWh for unplanned outages.

Value of Electricity: Based on a survey of Swedish households
Hypothesis: Consumers are able to estimate how they value planned and unplanned outages
Assumptions: Based on a survey of consumers
Value of Electricity: Ranges between $1.26 and $19.16/kWh

ICF Consulting Study

ICF Consulting published an issue paper in 2003 wherein they estimated consumers' willingness to pay. To apply this estimated data to determine the cost of the 2003 North American blackout, they multiplied the average value of electricity for the affected customers (including residential, commercial, industrial, and others), by the preliminary data on the magnitude and duration of this blackout.

ICF Consulting assumed that the value of electricity to consumers (measured as their WTP to avoid outages) is approximately 100 times the retail price of electricity. They cited an analysis done on the 1977 outage in New York City that resulted in a loss of more than 5,000 MW and lasted for 25 hours estimated that the direct cost was about $0.66/kWh (for example, losses due to spoilage, and lost production and wages), and an indirect cost of $3.45/kWh (due to the secondary effects of the direct costs).* Thus, the total unit cost of that blackout was $4.11/kWh or over $4,000/MWh in 1977. The national average retail price of electricity in 1977 for all customers was about $34/MWh.†

Value of Electricity: Study based on the impact of the 1977 New York City blackout
Hypothesis: New York blackout costs can be used to estimate WTP
Assumptions: Based on estimated damages and costs in 1977
Value of Electricity: $16.15/kWh (2014)

*See U.S. Congress, Office of Technology Assessment, "Physical Vulnerability of Electric System to Natural Disasters and Sabotage," OTA-E-453. Washington, DC, U.S. GPO, June 1990.
†See Annual Energy Review 2001, Table 8.6 available at www.eia.doe.gov/emeu/aer/elect.html.

ICF calculated initial estimates of the economic costs of this outage based on these ratios. Instead of providing only a point estimate for the total cost, they defined a range that is 80 times and 120 times the appropriate retail electricity price for the lower and upper bounds, respectively. Since there is considerable seasonal and regional variation in electricity prices, they used the August 2002 average electricity price of $93/MWh for the affected region (provided by the Energy Information Administration) to calculate the value of electricity to the customers affected by that outage.

ICF cited reports by the North American Electric Reliability Council (NERC) that the initial blackout that started around 4:00 p.m. (EST) on August 14, 2003, resulted in a loss of 61,800 MW and affected more than 50 million people. NERC also reported that by 11:00 a.m. (EST) on August 15, about 48,600 MW of lost power was restored, leaving a residual loss of about 13,200 MW. Finally, at 11:00 a.m. (EST) on August 16, NERC announced that "virtually all customers have been returned to electricity service."*

ICF assumed that the initial outage of 61,800 MW lasted for 4 hours and then half of that was restored, with the other half (30,900 MW) being the shortfall for another 10 hours. ICF assumed that another one half of the unserved 30,900 MW was restored after 14 hours and the remaining loss of 15,450 MW lasted for the subsequent 4 hours totaling 18 hours for the first phase of the blackout. Using similar arguments for the remaining period of the blackout, they constructed the estimates given in Table 7-2.

Using this scenario and the average electricity price for the affected region, the economic cost of this outage was estimated to be between $7 and $10 billion for the national economy.

Value of Service Estimates Based on Willingness to Pay

Willingness to pay estimates are based on the premise that a customer would be willing to pay to avoid the surprise loss of electricity service that results from an outage. Since the amount

*See a series of press releases and media briefings available at the NERC web site at www.nerc.com.

Table 7-2. 2003 Blackout Costs (ICF 2003)

Approx. Start Time	Approx. End Time	Lost Megawatt (MW)	Duration (hour)	MWH	Cost of Blackout ($ billions)	
					Lower Bound	Upper Bound
8/14 – 4 PM	8/14 – 8 PM	61,800	4	247,200	$1.8	$2.8
8/14 – 8 PM	8/15 – 6 AM	30,900	10	309,000	$2.3	$3.4
8/15 – 6 AM	8/15 – 10 AM	15,450	4	61,800	$0.5	$0.7
8/15 – 10 AM	8/16 – 12 AM	13,200	14	184,800	$1.4	$2.1
8:16 – 12 AM	8/16 – 10 AM	6,600	10	66,000	$0.5	$0.7
8/16 – 10 AM	8/17 – 6 AM	2,000	20	40,000	$0.3	$0.4
8/17 – 6 AM	8/17 – 4 PM	1,000	10	10,000	$0.1	$0.1
Total Economic Cost					**$6.8**	**$10.3**

the customer would pay for the lost energy is a direct measure of its value to him, willingness to pay can be used to measure the cost of an outage. In general, the value that customers receive from electricity is greater than or equal to the price they pay. If it were not, presumably, customers would not buy it. Therefore, the social cost of not delivering energy, including the possible inconvenience associated with an outage, is usually higher than its selling price.

Customer value of service (VOS) was calculated by DOE as a function of outage duration using a model from Sullivan, et al. (2009). Sullivan provides original VOS estimates for various customer groups using data from 28 consumer surveys conducted by 10 major electric companies between 1989 and 2005. These surveys assessed the cost of power outages to residential customers and commercial/industrial customers. Commercial and industrial customers were surveyed using the direct cost method. Each firm was asked to estimate the cost of hypothetical power interruptions. Residential customers were asked to report their willingness to pay to avoid outages. The WTP method was used to estimate the cost to residential customers because a substantial fraction of foregone consumer welfare does not translate into direct costs borne by residents.

DOE calculated outage costs using the two sets of VOS estimates derived using Sullivan, et al. (2009). An average cost function was determined for U.S. electric customers based on the total cost of each outage across 10 years, the average annual cost of outages caused by weather ranges from $18 to $33 billion.

The estimated costs by year are provided in the following figure and table. Costs range from $5 to $10 billion in 2007 to $40 to $75 billion in 2008. Outage costs due to hurricanes in 2008 are estimated to be $24 to $45 billion, while outages costs due to super storm Sandy in 2012 are estimated to be $14 to $26 billion.

These estimates account for numerous costs associated with power outages including lost output and wages, spoiled inventory, inconvenience, and the cost of industrial operations.

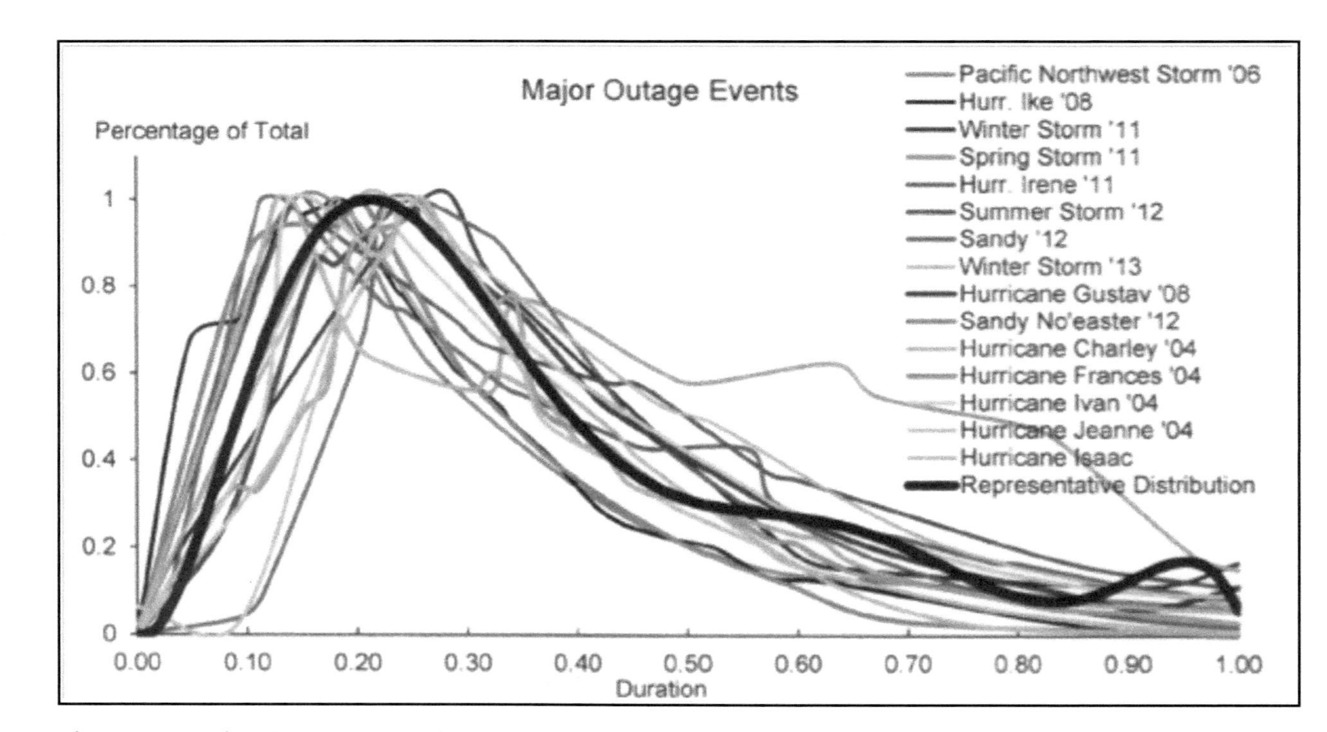

Figure 7 3. Major Outage Events (Department of Energy, Office of Electricity Delivery and Energy Reliability, OE-47)

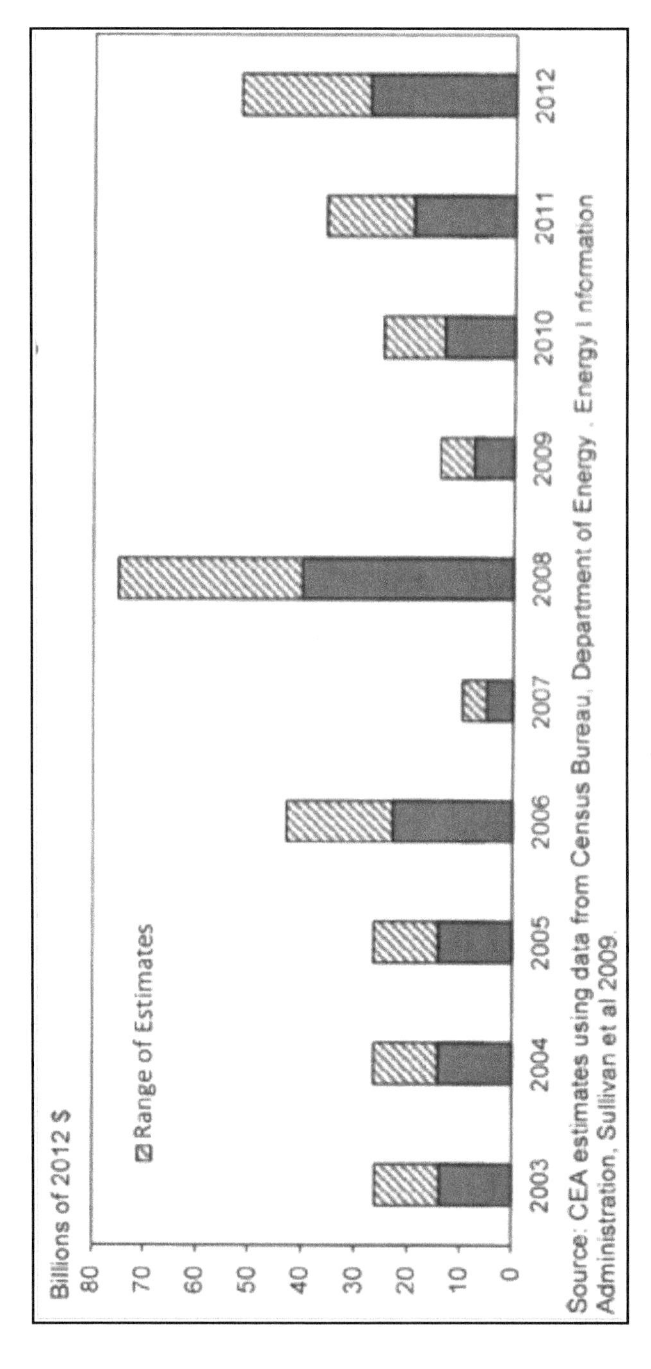

Figure 7-4. Estimated Costs of Weather-related Power Outages (CEA 2009)

Table 7-3. Estimated Costs of Weather Related Outages (Executive Office of the President 2013)

Year	Estimated Cost of Weather-Related Outages in the USA (billions 2012 $)
2012	$27 - $52
2011	$19 - $36
2010	$13 - $25
2009	$8 - $15
2008	$40 - $75
2007	$5 - $10
2006	$23 - $43
2005	$14 - $27
2004	$14 - $27
2003	$14 - $26

Value of Electricity: Office of the President estimates of weather related outages

Hypothesis: Severe weather events cause outages which, in turn impose economic hardship

Assumptions: EIA and other data

Value of Electricity: Annual impact up to $75 billion (2012)

STUDIES ON DEMAND-SIDE MANAGEMENT

Demand-side management strategies will affect the reliability of service as well as direct costs of service. The effects, as measured in terms of unserved energy, are potentially quite large. To illustrate the magnitude of the effects, Table 7-4 summarizes the results of a demand-side management analysis. The results, which apply to a 4500-MW peak-load system, illustrate the effect of various demand-side management strategies ranging from ones which reduce system peak demand by about 4% over a 15-year period without substantially changing total energy supplied to others, which employ demand-side management, plant post-plan or a combination of both. The amount of unserved energy is decreased if the demand-side management is implemented without changing capacity expansion plans. Unserved energy more than doubles if the strategy involves demand-side management and plant postponements. If the average cost of the unserved energy were $1.00 per kilowatt hour, the change in outage costs would be on the order of tens of millions of dollars per year. (EPRI P-4463-SR)

Table 7-4. Illustrative Impacts of Demand-side Management on Generation System Reliability costs in unserved energy for a 4500 MW peak load system (EPRI P-4463-SR)

Year	Case 1	Case 2	Case 3	Case 4
Year of Installation	1060	526	2494	2994
Installation +5	402	147	2995	7542
Installation +10	3253	1594	5846	21171

Two types of outages occur: unplanned and planned, or put another way: unintentional and intentional. The unplanned outages are the surprises that occur because of system unbalance, equipment failures or unexpected demands. Service interruptions are a part of many demand-side management programs. The interruptions range from air conditioning cycling to the direct shutdown of large industrial loads during periods when the electrical energy system is particularly stressed. The impact of these outages differs by customer class. The impact is also a function of timing, frequency, duration, and prior notification.

A number of studies have focused on outage costs or the cost to the customer not meeting their energy requirements. For the most part, these analyses can be classified as those that quantify outage costs based on the concept of consumer willingness to pay, those that base cost estimates on actual experienced losses (or prospective losses) by customers, and those that quantify outage costs based on the assumption that gross output is reduced in direct proportion to the amount of unserved energy. The first two approaches are closely related in that willingness to pay is correlated with losses in many instances.

A number of historic studies support these results. In 1975, Ontario Hydro surveyed customers in different use classes to estimate their willingness to pay to avoid sudden, total power outages of various durations. Table 7-5 shows the results of the study, which differentiate among large manufacturers, small manufacturers, commercial customers, and residential customers adjusted to 2014 dollars. Table 7-6 shows that the cost to manufacturers of an outage, especially one of short duration, is substantially greater than the cost to either commercial establishments or residences. Additionally, the cost per unit of unserved energy decreases as the duration of the outage increases, although the total cost of the outage increases. (EPRI P-4463-SR)

Given the Ontario Hydro outage cost estimates and estimates of outage duration distribution, we may calculate an average outage cost for each customer class. To compute this number for capacity expansion studies, EPRI made an estimate of the dis-

Table 7-5. Ontario Hydro Estimates of Outage Costs (2014 $/kWh)* (EPRI P-4463-SR)

Outage	Customer Class ($/Kwh)			
	Large Manufacturer	Small Manufacturer	Commercial	Residential
1 minute	150.56	213.31	5.01	0.43
20 minutes	22.57	34.74	4.25	0.39
1 hour	11.15	18.34	4.30	0.13
2 hours	9.61	18.84	6.47	0.09
4 hours	4.80	20.83	5.38	0.09
8 hours	4.60	16.45	4.84	0.04
16 hours	3.71	12.72	4.49	0.04

*Costs are translated from those in the Ontario Hydro study to 2014 dollars.

tribution of outages consistent with the average outage duration obtained from a detailed operations simulation of a typical utility. Table 7-6 presents the results of this calculation.

Table 7-6. DFI Estimates of Outage Costs by Customer Group Based Upon Ontario Hydro Study (EPRI P-4463-SR)

Customer Class		Outage Cost (2014 $/kWh)*
Large manufacturers		5.58
Small manufacturers		16.34
Commercial		5.08
Residential		0.30

*Residential outage cost includes both the amount customers would be willing to pay to avoid the outage and the cost of power. Costs are translated from those in the Ontario Hydro study to 2014 dollars.

Given these outage cost estimates for each customer class, a computation of the average cost of unserved energy for utility system customers as a whole was made taking the energy-weighted average of the customer class outage costs. Table 7-7 presents this calculation.

Value of Electricity: Based on DFI study
Hypothesis: Estimates developed by DFI using the Ontario Hydro data
Assumptions: Overall refinements in estimates by further analysis
Value of Electricity: Ranges between $.30 and $16.34/kWh

Cost of an outage can be estimated by the percent contribution of the class to peak demand in the system, rather than weighting by the percent of total energy demanded by each class. Table 7-8 presents the results of such a calculation.

EPRI has sponsored case studies to measure the actual costs associated with two outage incidents. One was a sudden outage of a few hours that occurred in San Diego, California. The other was a rotating blackout situation in Key West, Florida, which

Table 7-7. Cost of Outage Aggregated Over Customers for Hypothetical Utility (2014 dollars) (EPRI P-4463-SR)

Class	Fraction of Energy Sales in Class	Class Outage Cost ($/kWh)	Contribution to Average Outage Cost ($/kWh)
Residential	0.34	0.30	0.10
Commercial	0.39	5.08	1.98
Small manufacturers	0.135	16.34	2.21
Large manufacturers	0.135	5.51	7.42
		Average Outage Cost	5.03/kWh

Table 7-8. Cost of Outage Aggregated Over Customers for Hypothetical Utility Weighting Customer Classes by Fraction of Peak Demand (2014 dollars) (EPRI P-4463-SR)

Class	Fraction of Energy Sales in Class	Class Outage Cost ($/kWh)	Contribution to Average Outage Cost ($/kWh)
Residential	0.45	0.30	0.14
Commercial	0.36	5.08	1.83
Small manufacturers	0.10	16.34	1.63
Large manufacturers	0.09	5.51	0.49
Average Outage Cost			4.08/kWh

occurred over a 26-day period. Tables 7-9 and 7-10 show some results of these studies. (EPRI P-4463-SR).

In each case researchers were dispatched to the local shortly after due outages so as to obtain survey data which were relatively unbiased.

Numerically, they are similar to results derived from the Ontario Hydro surveys. However, the EPRI estimates are based on estimates of actual costs incurred for specific real outages. The EPRI approach avoids a major difficulty faced in estimating

Table 7-9. Shortage Costs in San Diego Shortage (2014 dollars) (EPRI P-4463-SR)

	Industrial Customer ($/kWh)	Commercial Customer ($/kWh)
Direct User	9.68	8.31
Employees of Direct User	0.74	0.30
Indirect User	0.41	0.46
Total	10.83	9.07

Table 7-10. Shortage Level and Short-Run Shortage Costs: Key West (2014 dollars) (EPRI P-4463-SR)

	Shortage Level	Shortage Cost (2014$/kWh)
Non-residential Users	4.8% for 26 days	
Producer Costs		6.84/kWh
Employee Losses		0.35/kWh
Consumer Losses		0.69/kWh
Total		7.88/kWh
Residential Users	7% for 26 days	
Comfort and Convenience		$0.17/kWh
Fires and Other Losses		Insignificant

shortage costs; that is, first impressions tend to exaggerate shortage costs because they invariably overlook the intricate coping devices that can greatly mitigate shortage impacts. For example, if production is shut down because of power outage, it can frequently be made up later simply by working overtime. Thus, the cost to direct customers may be the labor premium for overtime, plus any inconvenience, not the value of the product as one might initially estimate

Table 7-11 depicts the result of another estimate of average customer interruption costs.

Table 7-11. Estimated Average Customer Interruption Costs U.S. 2014 Dollars by Customer Type and Duration (Sullivan, et al., 2009)

<table>
<tr><th rowspan="2">Customer Type</th><th rowspan="2">Interruption Cost Summer Weekday</th><th colspan="7">Interrupted Duration</th></tr>
<tr><th colspan="3">Monetary</th><th>30 min</th><th>1 hr.</th><th>4 hr.</th><th>8 hr.</th></tr>
<tr><td>Large C&I</td><td>Cost per Average kWh</td><td colspan="3">$192</td><td>$42</td><td>$28</td><td>$20</td><td>$16</td></tr>
<tr><td>Small C&I</td><td>Cost per Average kWh</td><td colspan="3">$2,641</td><td>$617</td><td>$414</td><td>$338</td><td>$299</td></tr>
<tr><td>Residential</td><td>Cost per Average kWh</td><td>$24</td><td>$5</td><td>$3</td><td>$1</td><td colspan="3">$1</td></tr>
</table>

Value of Electricity: Based on literature searches
Hypothesis: Various surveys and analysis assumptions. Surveys and analysis are useful in estimating outage costs.
Value of Electricity: Ranges between $.15 and $39.401/kWh

An extensive review of the interruption cost literature both in the United States and abroad has been conducted by Sanghvi (39, pp. 8-1 to 8-50). Sanghvi reviewed the various methodological approaches in order to credibly compare estimates from different sources and posit his own "order of magnitude" range of value for the residential, commercial and industrial sectors. Converted to 2014 dollars, Sanghvi's estimates are:

Residential $0.15/kWh to $4.20/kWh
Commercial $2.81/kWh to $39.40/kWh
Industrial $2.81/kWh to $19.61/kWh

Value of Electricity: Based on surveys conducted immediately after blackouts in two cities
Hypothesis: Consumers could accurately determine actually losses
Assumptions: Based on surveys
Value of Electricity: Ranges between $.17 and $7.88/kWh

Chapter 8

Other Methods to Estimate Value

USING UTILITY RESTORATION COSTS AS A MEASURE OF VALUE

When large storms or other disasters damage electric systems, utilities launch massive round-the-clock efforts to restore power quickly. The logistics associated with these restoration efforts are daunting. In addition to deploying their own crews, utilities call upon crews from other parts of the country and often Canada to help, with the "host utility" paying for supplies, wages, equipment rental, transportation, hotel rooms, meals, and even laundry. Added to that are equipment costs, miles of new wire, thousands of new poles, new transformers, cross arms, fuses, and many other costs. (EEI 2005)

Utilities incur substantial costs to repair their systems after disasters strike. Based on survey data obtained by the Edison Electric Institute (EEI) for 81 major storms from 14 utility respondents, these disasters cost utilities approximately $2.7 billion (in constant $2003) between 1994 and 2004). However, the economic impact of not having electric service in an area hit by a disaster is much larger than the cost of repairing the damage. This suggests that the utilities' current practice of incurring additional costs to mobilize outside resources to restore power quickly is appropriate. In at least one instance, Wall Street changed its credit outlook for a utility, in part because of concerns over how quickly a decision favorable to the utility would be reached to mitigate the financial impact of restoration expenses.

There is little consistency in establishing which events do, or do not, qualify for disaster mitigation. For example, one company was required to expense approximately $160 million of O&M

storm costs associated with a major hurricane against current year earnings, while another utility was allowed to recover a $1 million storm expense over a four-year period.

Storm reserves provide a type of self-insurance to pay for major storms; however, they may not be funded sufficiently to pay for catastrophic storms. In most instances, these reserves do not provide a ready source of cash to pay for storms. When faced with significant O&M restoration costs that could require a substantial write-off, many companies are granted permission by their commissions to defer these costs, but there is often a lengthy delay in providing this relief and the approval process can become politicized.

For example, over a six-week period beginning August 13, 2004, four hurricanes struck Florida. Never before in the state's history had so many hurricanes hit in a single season. The scale of the destruction caused by the storms was also unprecedented, with one in five homes suffering damage. The impact on Florida's investor-owned electric utilities was equally destructive. The hurricanes required the state's investor-owned utilities to replace more than 3,000 miles of wire, almost 32,000 poles, and more than 22,000 transformers. (See Table 8-1)

To gauge the potential financial impact of major storms, EEI examined the impact that very large storms occurring between 2000 and 2004 had on four utilities, all Investor Owned Utilities (IOUs). Table 8-2 evaluates company transmission and distribution (T&D) expenses and net earnings using data from media accounts of storm costs and Federal Energy Regulatory Commission (FERC) Form 1 financial data to compare the cost (including capital) of four large storms. The data indicates that storm costs can have a large and potentially devastating financial impact. In some instances, storm costs exceed a company's total earnings and T&D expenses for the entire year.

Local blackouts are not unique to U.S. cities as is evidenced by a report regarding the U.K. community of Salisbury. (Ross 2013) Salisbury experienced 6 power cuts in 11 days in 2013. A fault in an underground cable that the power company was unable to lo-

Table 8-1. Florida 2004 Hurricane Damage* (EEI 2005)

	Poles Replaced	Transformers Replaced	New Conductor (Miles)
Hurricane Charley			
FPL	7,100	5,100	900
Progress Energy	3,820	1,880	667
Hurricane Frances			
FPL	3,800	3,000	550
Progress Energy	2,800	1,560	500
Hurricane Ivan			
Progress Energy	100	570	N/A
Gulf Power	5,060	3,175	225
Hurricane Jeanne			
FPL	2,300	3,000	250
Progress Energy	6.720	4,010	100
Total	31,700	22,295	3,192

Source: Company reports
*Comparable storm damage data for Tampa Electric is not available

Table 8-2. Impacts of Large Storms 2000-2004 (EEI 2005)

Storm Description	Date	Storm Cost $Million ($2003)	Financial Impact	
			$ of Annual T&D Expenses	% of Net Operating Income
Progress Energy – NC Ice Storms	2000	$205	259.8%	96.7%
Dominion Energy – Hurricane Isabel	2003	$212	72.3%	24.8%
Progress Energy – Florida Hurricanes	2004	$366	303.8%	104.1%
FPL – Hurricanes	2004	$890	305.2%	97.0%

Source: EEI 2005 from Press Accounts and FERC Form 1 Data

cate meant shops and restaurants lost thousands and faced even greater losses at the busiest time of the year. For example, Fergus McMurray, general manager at Côte restaurant, estimated their losses at about £9,000. The restaurant Strada had to turn away a party of 21, losing £900 from one table. They estimated they lost up to £6,000.

Some shops in Salisbury now have generators, cables and lights so they can stay open, while staff at some shops like the Silver Gift Gallery have been forced to work in the dark and only accept cash. And at Michael's Hair Shop, the power cuts mean no hot water to wash people's hair, no lights to see what they're cutting, and no power for blow dryers.

VALUE OF UNSERVED ENERGY

Utility planners have for years used figures in the range of $1.00 to $2.00 per unserved kWh in their planning studies. In these cases, the planner would design a power system aimed at a certain targeted loss-of-load probability, often in the order of one day in ten years. In conducting the analysis of various system configurations, the planner would assign a cost of $2.00/kWh to the loss of all load for one day in ten years.

THE VALUE OF ELECTRIC LIGHTING

A Ph.D. student at Carnegie Mellon University (Schnitzer, 2014) studied quality alternatives fossil-fueled lighting in several African countries. He derived a functional form for the willingness to pay for grid-supplies electric lighting. He determined that electric lighting saves consumers on the order of 1 to 5 dollars per month and increases consumer surplus arrange between 2 and 18 dollars per month.

IEA continues to use a supply oriented definition of access for their modeling exercises, and defines consumption levels in

Watts and kWh that are aligned with service "tiers" of access to electricity. In IEA modeling the levels of consumptions specified as "access" for rural households is 250 kWh per year using an average of approximately 35 Year/year and a benefit of 10 dollars per month that translate to a value of electricity of $3.43/kWh.

Using a strict comparison between light sources and assuming kerosene lamps have an output of 10 lumens and CFL 450 lumens, Schnitzer valued the electric lighting at $0.22/kWh.

RESILIENCY

A more resilient grid is one that is better able to sustain and recover from adverse events like severe weather—a more reliable grid is one with fewer and shorter power interruptions felt by the customer. The benefits from microgrids during Sandy were unique to those few customers served by the microgrid. For the remaining and majority of customers impacted by Sandy, there are other number of beneficial improvements to distribution resiliency which could be employed in storm-prone areas.

Resiliency strategies are increasingly debating microgrids that provide electric service capacity when the grid is not available. As an illustration, in Connecticut, the state developed a Microgrid Grant and Loan Pilot Program to develop microgrid solutions that can provide power to critical facilities. The state is investing $1.5 million upfront to fund preliminary design and engineering costs for selected finalists. The state will invest an additional $13.5 million for microgrid projects.

Cases where a local microgrid survives a bad storm are not unusual. And, they are receiving a lot of press coverage. Recent publications cite benefits, challenges and investments in microgrid powering options. *IEEE Power Energy Magazine* devoted its July/August 2013 issue to the topic opening with the marquee "Microgrids—Coming to a Neighborhood Near You." There has also been favorable legislation for distributed generation technologies, such as in California and Connecticut. Combined heat and

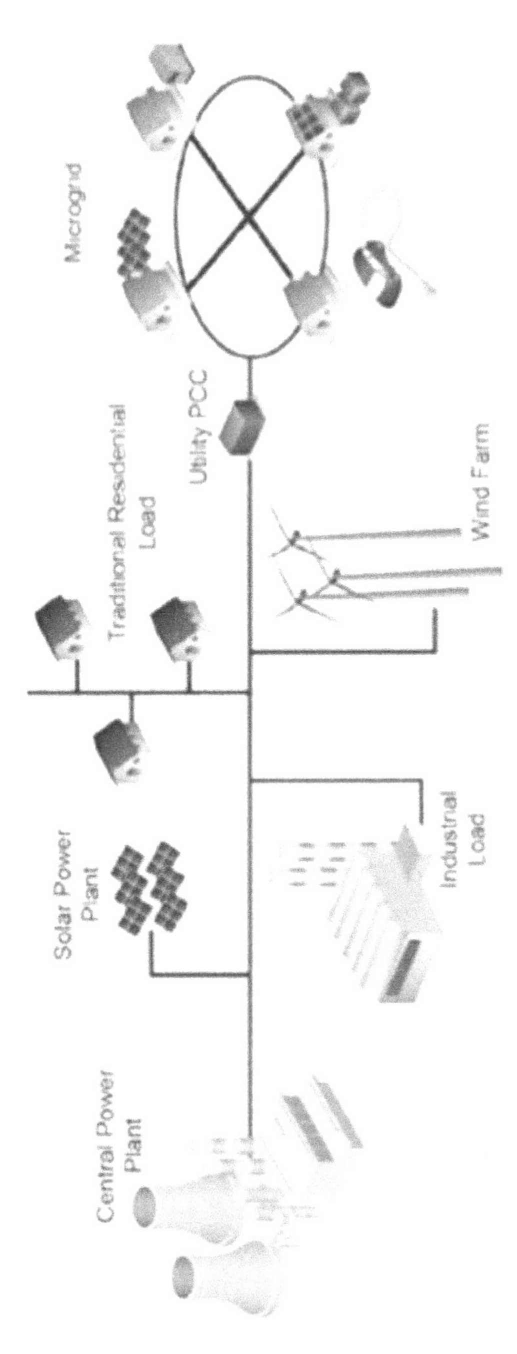

Figure 8-1. Microgrid as Part of a Traditional Utility System (IEEE 2013)

power (CHP) is the subject of a 2012 Obama Executive Order to double U.S. CHP capacity by 2020. All of these signs persuade a closer look at the microgrid concept as part of developing a more resilient and flexible electric grid in the future.

Whether or not these distributed resources become a large part of the future grid, the idea of an integrated approach and of capturing the value of many resources, has merit. This concept is inherent to the operation of the existing power grid, which offers interconnectivity as an intrinsic attribute. Microgrids follow along many of the same technical arguments. Properly integrated, with operational collaboration and utility knowhow, they will likely contribute to flexibility and resiliency needed to meet the demand for continuous supply of electric power in the future. (EPRI Microgrids 2013)

The Costs of Independent Systems

EPRI recently estimated the value of providing reliable electric service with an independent electric system using local generation and compared that to the cost of grid-supplied electricity. The study analyzed two topologies of local generation installations:

1. Local generation installations which augment connectivity to the grid.* In this topology, the local generation system may provide much or all of the building's needs at some times—but rely on the grid for electrical "support."

2. Local generation installations which are not connected to the grid. In this topology, the building may or may not aspire to provide the same level of service as a grid-connected consumer.

Figure 8-2 illustrates these two topologies using photovoltaics (PV) systems as an example.

---

*The grid in this monograph is assumed to be the entire power delivery infrastructure from connection to bulk power generation to transmission and distribution.

From a societal perspective, local generation installations which are connected to the grid offer the lowest overall cost!

The author conducted an analysis of the value grid support provides to consumers who install local generation. The analysis assumes that consumers with local generation who remain connected to the grid will continue to receive grid support for no special connectivity cost beyond what a traditional non-connectivity consumer pays. In order to value what grid connectivity is worth, an analysis was performed assuming that the consumer would add the technologies which would provide the same functionality as the grid provides.

Figure 8-3 illustrates the estimated average cost of the grid to the average residential customer broken down into transmission and distribution charges using data for the average U.S. residential consumer.

These averages do not truly reflect the large range of payments from residential customers of varying sizes as seen in Figure 8-4.

Figure 8-5 illustrates a comparison of consumers who are not employing local generation with those who do use local generation. The analysis illustrated is based on the assumption that there is modest overall penetration of local generation. Note that the average monthly payment for grid services drops from $40 per month to $20 per month despite the fact that the same relative services are being provided.

The analysis changes significantly as the penetration of local generation increases. Figure 8-5 illustrates this. As consumers without local generation are left with the burden of supporting the grid when there is a relatively high penetration of local generation (assume 30% of customers produce 50% of energy locally), the grid cost jumps to $48 per month, whereas without changes in rate design, the consumers with local generation are only paying $24 per monthly, clearly illustrating the subsidy between classes of customers who have local generation by those who do not.

For most of this analysis, costs and benefits are put in terms of cents/kWh such that 1 cent/kWh = $87.60/kWh/kW/year or 8760 times 1 cent.

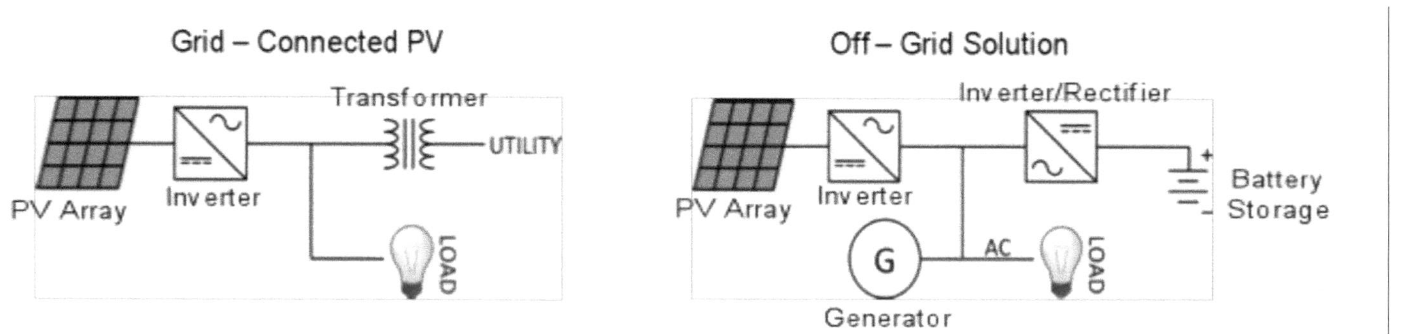

Figure 8-2. Alternative Topologies for Local Generation Installations

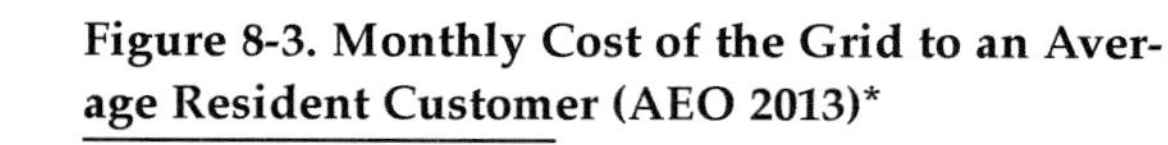

Figure 8-3. Monthly Cost of the Grid to an Average Resident Customer (AEO 2013)*

*AEO 2013 estimates of nominal cost for transmission = 1.1 ¢/kWh and for distribution = 3.1 ¢/kWh. Actually, totals $10.80 for transmission and $30.44 for distribution.

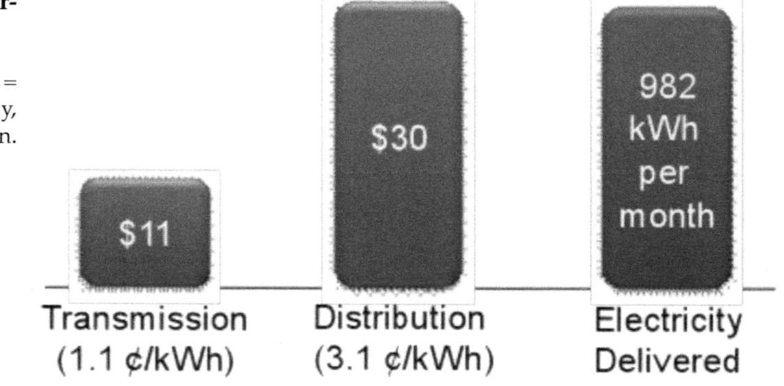

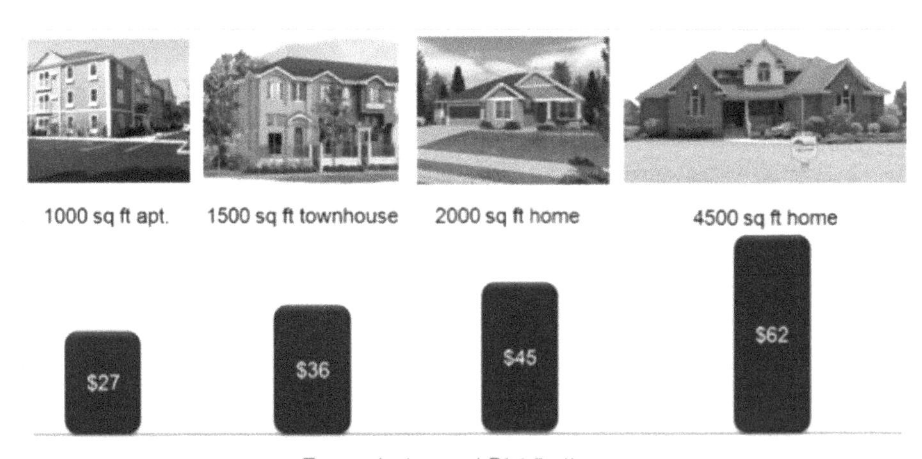

Figure 8-4. Ranges of Monthly Payment for the Grid (AEO-2013)

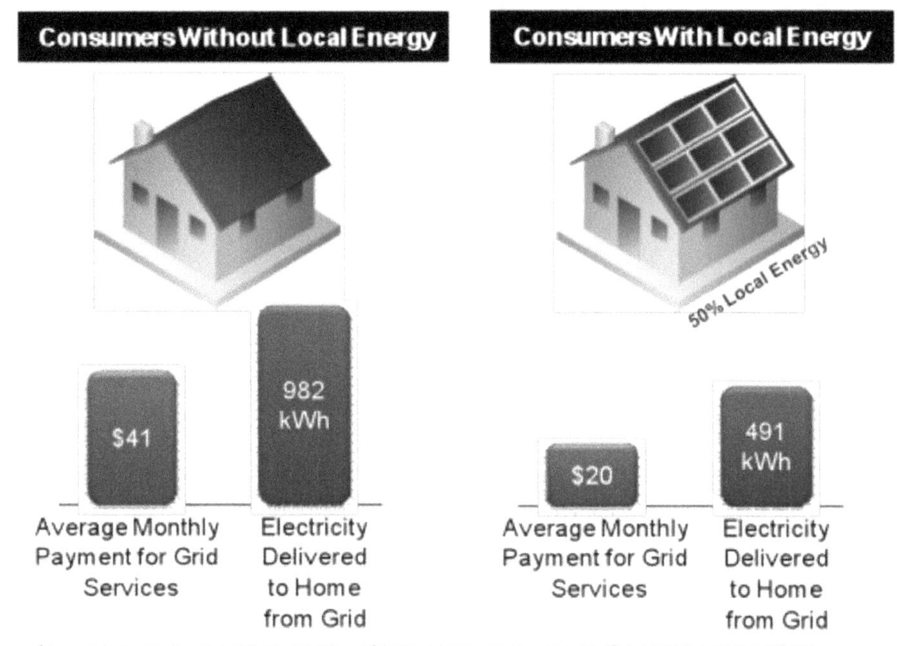

Figure 8-5. Costs to Customers with Local Energy Source for Grid Connectivity

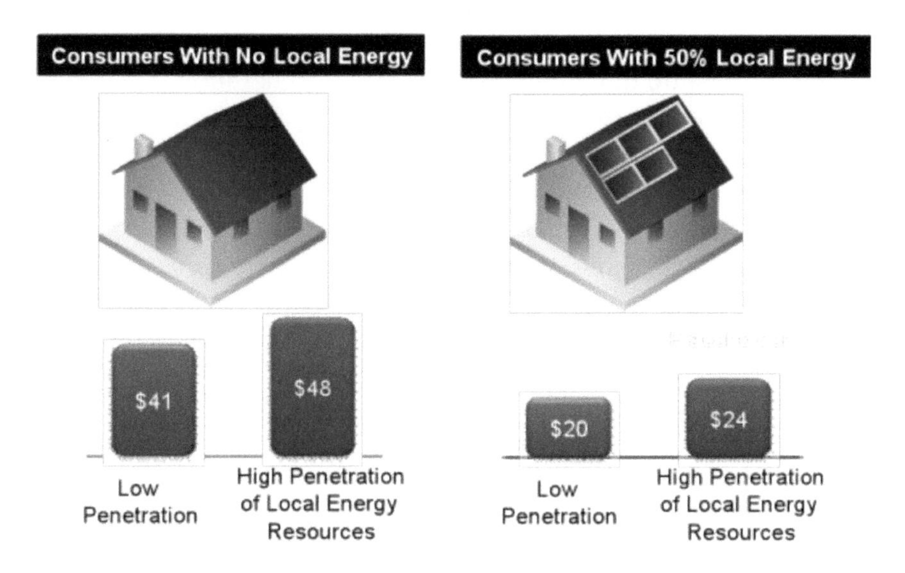

Figure 8-6. Costs to Customers for Grid Services with a High Penetration of Local Energy Resources

Therefore, if the average distribution cost is equal to 3 cents/kWh and if this were the distribution capacity change then this would equate to $262.80/kW-yr. Using an arbitrary* value of $25/kW-yr[18] for participation in demand response programs then this could be equated to the amount a generation provider can monetize using a modernized grid which provides for two-way communication and control and local resources used for grid services. Comparing this with a generation capacity charge,† it positions local generation as very competitive.

---

*One possible way to value what the contribution is for local generation is to review the price paid in wholesale power markets for peak load reduction. Programs vary, but the following are examples:

- The New York Power Authority (NYPA) pays customers $25/kW.
- PJM pays DR capacity the market clearing price ($/MW-day) for the zone in which the premise resides. To get equivalence, multiply the PJM price by 365. The recent auction in PJM cleared at price of $50-120/MW-day, $18-43/kW-year.

†In the latest PJM auction, capacity cleared at ≈ $50/kW-yr while the nominal value is $100/kW-yr. This is consistent with the fact that 80% of PJM generation costs are related to energy, 15% to capacity, and 5% to ancillary services.

The original estimates of \$40/month grid cost and \$70/month generation cost do not adequately reflect the generation cost which is actually 80% energy, 20% capacity and ancillary services. Careful examination of these costs conclude that volumetric approach to rate design based on kWh consumption will not recover the cost of the grid in the case of consumers with local generation.

Further costs and benefits which need to be considered include those which apply to consumers with local generation. These may include additional metering cost, integration cost, and the reduction in losses.

Technology Uncertainties in Distributed Power

These estimates are for a residential case using the most likely technologies that will be used for an off-grid power system. Other technologies such as an independent campus microgrid will have different cost implications; however, the central conclusion that an independent microgrid does not provide consumers the reliability of the grid and other services without adding substantial technology and cost.

As indicated above, in order to estimate the value of the grid, the author compared two houses. One is a house which remains grid-connected but has installed photovoltaic power generation such that the consumer is able to operate the home on an annual net-zero energy basis. This house has an 8 kW solar array which produces approximately 11,800 kWh/year. The installation will cost approximately \$48,000.

For the second system, the author took the same home and assumed that there was no grid connection and estimated the cost of the added technologies which would be required to provide electric service approaching that supplied by the grid. The added technologies included the following:

Increasing the PV array size from 8 kW to 11 kW	\$18,000
72 lead acid batteries	36,000
Battery charge controller	5,400
Standby Generator (14 kW)	4,000
Total	**\$63,400**

If consumers elect to install these technologies in order to replicate the same level of service as the grid provides, they would incur an additional capital cost of $63,400. Assuming amortization at 6% over 20 years, this translates to a monthly cost of $272 (see Figure 8-7).

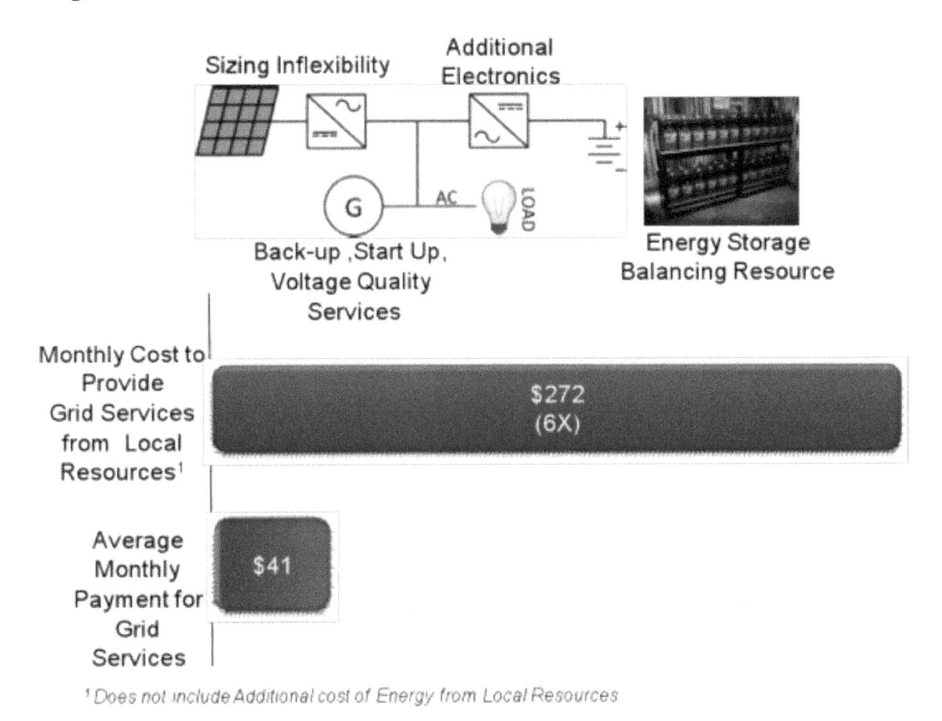

Figure 8-7. Cost if Grid Services are provided by Local Energy Resources

Figure 8-8 illustrates the cost of providing grid services from local resources using the St. Louis, Missouri, area.

ARRA RESULTS

The U.S. Department of Energy (DOE) made several observations of the value of service from improvements in reliability from smart grid projects funded by the American Recovery and Reinvestment Act (ARRA). One project involved 230 automated

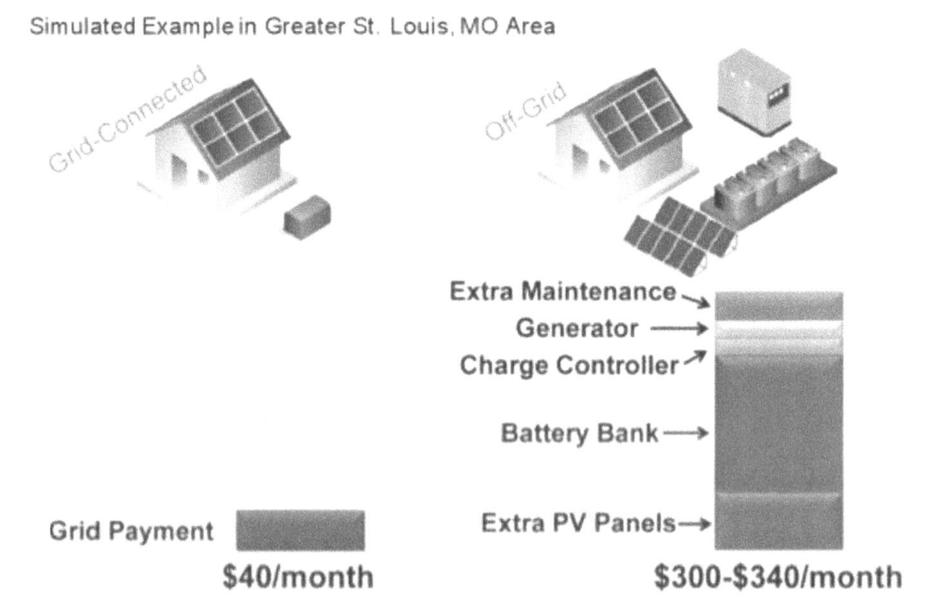

Figure 8-8. Cost of Grid Services from Local Resources

feeder switches on 75 circuits in an urban area during the period of April 1 to September 30, 2011. During this period in 2008, System Average Interruption Duration Index (SAIDI) improved 24%; the average duration decreased from 72.3 to 54.6 minutes—a total of 17.7 minutes.

The estimated monetary value of this improvement in reliability based on value-of-service data is $21 million.

One example from DOE's projects involves the response from a July 5, 2012, storm at the Electric Power Board of Chattanooga (EPB) shown in Figure 8-9. They estimated that the smart grid technologies collectively improved storm response by 17 hours offering a $1.4 million cost reduction.

ERCOT ECONOMICALLY OPTIMAL RESERVE MARGIN

In 2014, The Brattle Group published a study which estimated the economically optimal reserve margins for the Electric Reli-

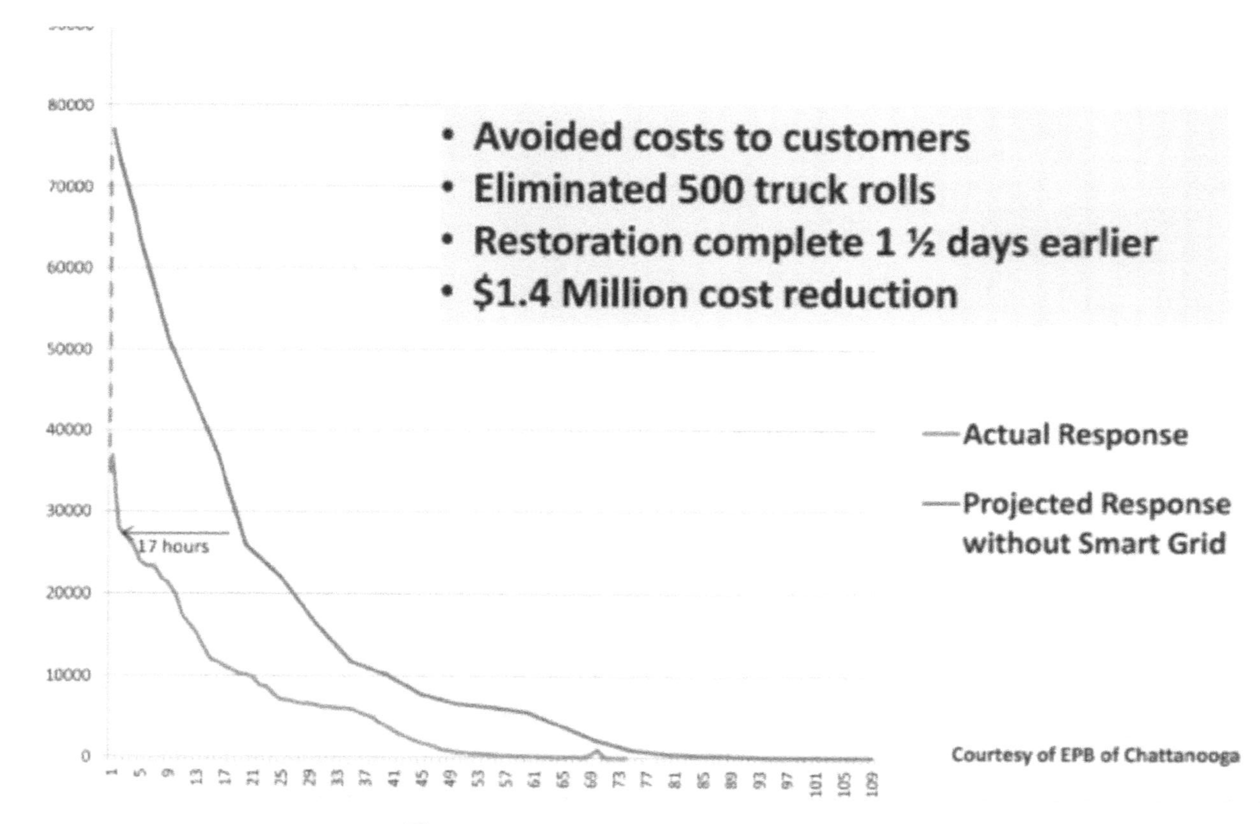

Figure 8-9. July 5, 2012 Storm Response (EPB)

ability Council of Texas (ERCOT.) (Brattle 2004) The economically reserve margin is in itself an estimate of the value of electricity.

The best estimate of the value of electricity using these studies is to examine the value of non-controllable load resources that actively participate in ERCOT's response reserve service (RRS) market. These characteristics imply a marginal cost of electricity ranging up to an average of $2,569/MWh, ranging up to $9,000/MWh for price responsive demand response.

INSIGHTS FROM CAPACITY MARKETS

Reliability requirements have generally been interpreted to require the purchase of capacity in excess of expected loads by a reserve margin. The reserve margin, between 15 and 20% of expected peak load, is designed to ensure reliability under worst-case conditions including high demand, forced generation outages, and transmission outages. The reliability goal has generally been characterized as no more than one loss of load event in 10 years. (Bowring 2013)

Electricity markets do not prevent the possibility of blackouts, and it can be assumed that this will continue to be the case. Given the demand-side flaws, fully eliminating blackouts due to insufficient generation is unlikely to be optimal. To see this, define the "Value of Lousy Load" (VoLL) as the amount that consumers would pay to avoid having supply of power interrupted during the blackout. (IAEE 2013) Now suppose the average annual *Duration* of blackouts is five hours per year and that VoLL = $20,000/MWh. Suppose further that the rental cost of reliable capacity (RCC) is $80,000/MW-year. If one MW of capacity is added, it will run five hours per year on average and reduce the cost of blackouts by $100,000/year. That is more than the cost of capacity so new capacity should be built up to the point where the duration of blackouts falls to four hours per year and the marginal cost of capacity equals the marginal reduction in the cost of lost load. That is, the optimal expected duration of blackouts is *Durations* =

RCC/VoLL. As long as the rental cost of reliable capacity is positive, efficiency requires that blackouts occur with positive probability.

Key insight is that electricity markets cannot optimize blackouts. (IAEE 2013)

COMPARING HAVE AND HAVE-NOT SOCIETIES

An alternative method of estimating the value of electricity is to compare two societies—one which has adopted electricity and associated technology such as the western world (e.g., Japan, U.S. Canada and western Europe), and one which has not, such as India, portions of South Africa and Brazil. This is illustrated in Figure 8-10*. Note that 98.51% of Brazil households have electricity, 18.5% of South Africans do not. (Based on interviews of members of CIGRE's Steering Committee, Recife, Brazil, December 2014).

Figure 8-10. Determining the Value of Electricity: Conduct a Comparison

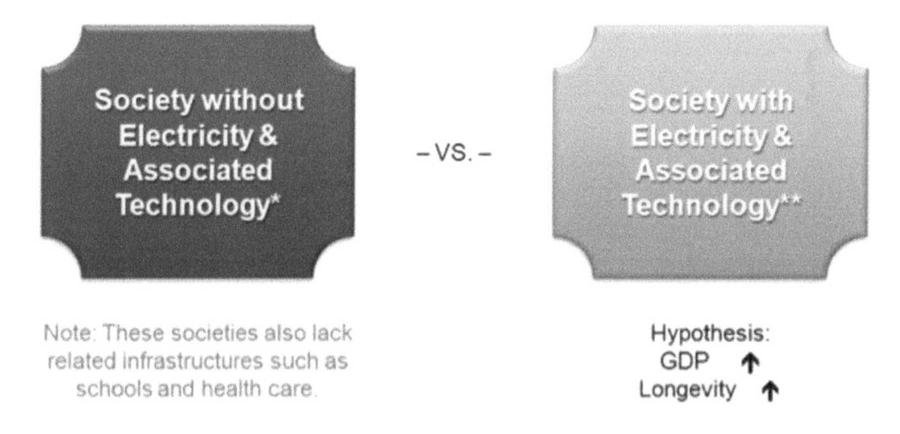

Chapter 9

Enhancing the Value of Electricity

WHAT CAN BE DONE TO ENHANCE ELECTRICITY'S VALUE TO CONSUMERS?

Strategically, the electric utility industry must take control of its own destiny and mold customers' perceptions: it must successfully educate customers about electricity's value and the role it plays in their daily lives. Electricity's value comes from its utility—the basic needs it fills (comfort from a warm room on a cold day), the tasks it performs (cooking food), the services it renders (money from an automated teller machine).

As illustrated in Table 9-1, utilities can enhance electricity's value in three ways:

- Increase customers' perceptions of electricity's value
- Promote electrification
- Focus on electricity's own technical and economic attributes
- Incorporate the concept of "enhanced value" into all planning processes

First, utilities can enhance electricity's value simply by increasing customers' *perceptions* of its value. They can discover what customers want and need and then show them how electricity meets these requirements. Utilities must recognize that customers do not perceive a need for electricity specifically—they *do* think they need heat, light, stereo systems, and "quick cash" machines, however. (Similarly, customers failed to perceive that they needed, or would use, personal computers until they became readily available and accessible.) If they increase customer satis-

Table 9-1. Determining the Value of Electricity: Conduct a Comparison

Value Enhancement	Outcome
Increase customer perceptions of value	Increasing perception is everything: real value may not be meaningful
Promote electrification	Generally gain support for applications of electricity which beneficially displace fossil fuel use
Enhance electricity's technical and economic attributed	Deliver more real value
Make value an integral part of meeting customer needs	Both increase perception and deliver value

faction, utilities can also enhance their own public images, which, in turn, impacts perceptions about electricity.

Commercial and industrial (C&I) enterprises, including public institutions, are facing intense pressure to increase productivity and reduce operating costs without compromising quality of production or service. In many cases, electrification—i.e. the application of novel, energy-efficient electric technologies as alternatives to fossil-fueled- or non-energized- processes—can boost productivity, reduce energy intensity, lower operating costs, and enhance the quality of service to the enterprise and the customers that it serves. In addition, many electrification applications may generate substantial emissions reductions that may benefit both the customer and the public. Examples of industrial C&I electrification applications include non-road electric vehicles and materials handling equipment, electric heat pumps, and electric process heating.

However, a lack of familiarity and experience with emerging technology impedes many enterprises, particularly small to medium-sized customers and public institutions, from pursuing electrification measures that may improve their productivity, and may enhance their quality, and have the potential for lower costs. Such enterprises may benefit from information and support from their electric utility.

Electric utilities themselves face obstacles to serving as effective partners in this regard. Identifying and measuring the prime opportunities for electrification in a given service territory can be elusive. Utilities have needed to reconcile electrification strategies with mandated energy efficiency goals that are usually narrowly defined in terms of kilowatt-hour reductions. Moreover, the lack of an analytical framework to quantify the net benefits of electrification strategies—from the customer-, utility- and the public's perspectives—hinders the development of utility-business partnerships to facilitate beneficial electrification.

Utilities can enhance the *technical capability* of electricity to do work. As end-use products (appliances/ devices) become more functional and meet customer needs more effectively, they become indispensable to customers' everyday lives. Each successive generation of electric-powered products should be able to "do it quicker, sooner, and more effectively." Utilities must link these improvements with electricity's role in facilitating them.

In addition, customers will not recognize the full potential of electricity's value until utilities deploy the most cost-effective programs/ services. Electricity's *economic value* will improve as utilities design and promote more effective demand-side management (DSM) programs that meet actual customer needs. Utilities that match customer needs to the right energy service programs will design and implement the most economic, cost-effective programs.

Finally, utilities that want to enhance electricity's value must consciously do so—they must incorporate this goal into all of their planning processes. Traditional utility planning focused solely on supply-side issues. This traditional process assumed stable and predictable load growth; "the customer" received very little consideration in this process. As the utility marketplace changed, utility planning processes had to evolve. Accordingly, utilities began adopting processes that focused more on demand-side issues, such as least-cost planning. However, traditional least-cost planning procedures concentrated on economic motivations and neglected other factors that may prompt customers to purchase utility services—such as perceptions of value. An integrated planning approach—both in resource and program planning—incorporates both economic and non-economic factors.

Many electric utilities are turning to a value-based philosophy when designing, implementing, and promoting programs and services. While many utilities may not have labeled their actions as such, the evidence mounts, including the manner in which they develop and market DSM programs and services.

Utilities now promote new technologies in a variety of innovative ways. Many utilities now sell, lease, install, and/or maintain equipment for a variety of end uses or do so in conjunction with third parties. (This is a controversial practice in many states, since contractors and others believe utilities are usurping their roles.)

Changes in Demand Offer Harbingers of Value

Looking toward the future, there are a number of changes in the demand for electricity from grid-related services which may act as barometers of value.

Electricity demand is driven by a complex combination of the desire for the comfort, convenience, and productivity that energy provides, but is tempered by the price, availability, and functionality of competing energy forms (e.g., natural gas) and the technology that converts electricity into services. Fundamental to this demand is the presence of income (in households) and/or economic activity (in business and industry). In each energy application, the technology that converts electricity or other energy forms into lighting, heating, cooling, motive power, or other energy services plays a key role in the effectiveness of the desired energy service, its economy, and the resultant environmental footprint.

Until recently, the relationship of electricity demand to these factors was straightforward. Now with the advent of photovoltaic power generation, the expanding use of combined heat and power (CHP), and dramatic changes in the end use of electricity, the equation changes. Consumer demand for electricity is now potentially supplied by a combination of grid-supplied energy services and power generated on site.

The advent of new and improved technology also has a substantial impact on the technology landscape and then the overall demand for energy. Recent adoption of plug-in electric vehicles (PEVs) is a prime example of new technologies that can significantly increase the use of electricity. Other new technologies such as tablets, PCs, home entertainment systems, and heat pumps are additional examples.

Figure 9-1 summarizes the key "puts and takes" of changing electricity demand relative to the 2012 Annual Energy Outlook. Each of these elements represents opportunities for value-added services which utilities could use to increase the value which electricity provides.

Figure 9-1 primarily influences demand from grid-related services.

The results indicate that each of the "puts and takes" indicated in the figure can have a substantial impact on the resulting demand for electricity. In each case, there is substantial uncertainty in the estimates provided. EPRI has identified each of these variables as key markers in tracking the demand for electricity from grid-related services. In this report, EPRI uses the Annual Energy Outlook (AEO) 2012 as the base case.

The Annual Energy Outlook

One starting statistic in reviewing AEO2012 is that the Annual Energy Outlook has been estimating a continuing decline in electricity consumption each year since 2007 (see Figure 9-2). As suggested, this reflects a consistent increase in the end-use energy efficiency of technology and buildings, despite dramatic growth in adoption of consumer electronic technology.

If deregulation becomes more widespread, there could be downward price pressure on electricity, which would increase demand somewhat. The low wholesale price of natural gas is not yet fully reflected in the retail price. When it is fully reflected, more fuel switching (electric to gas) may result. However, whole-

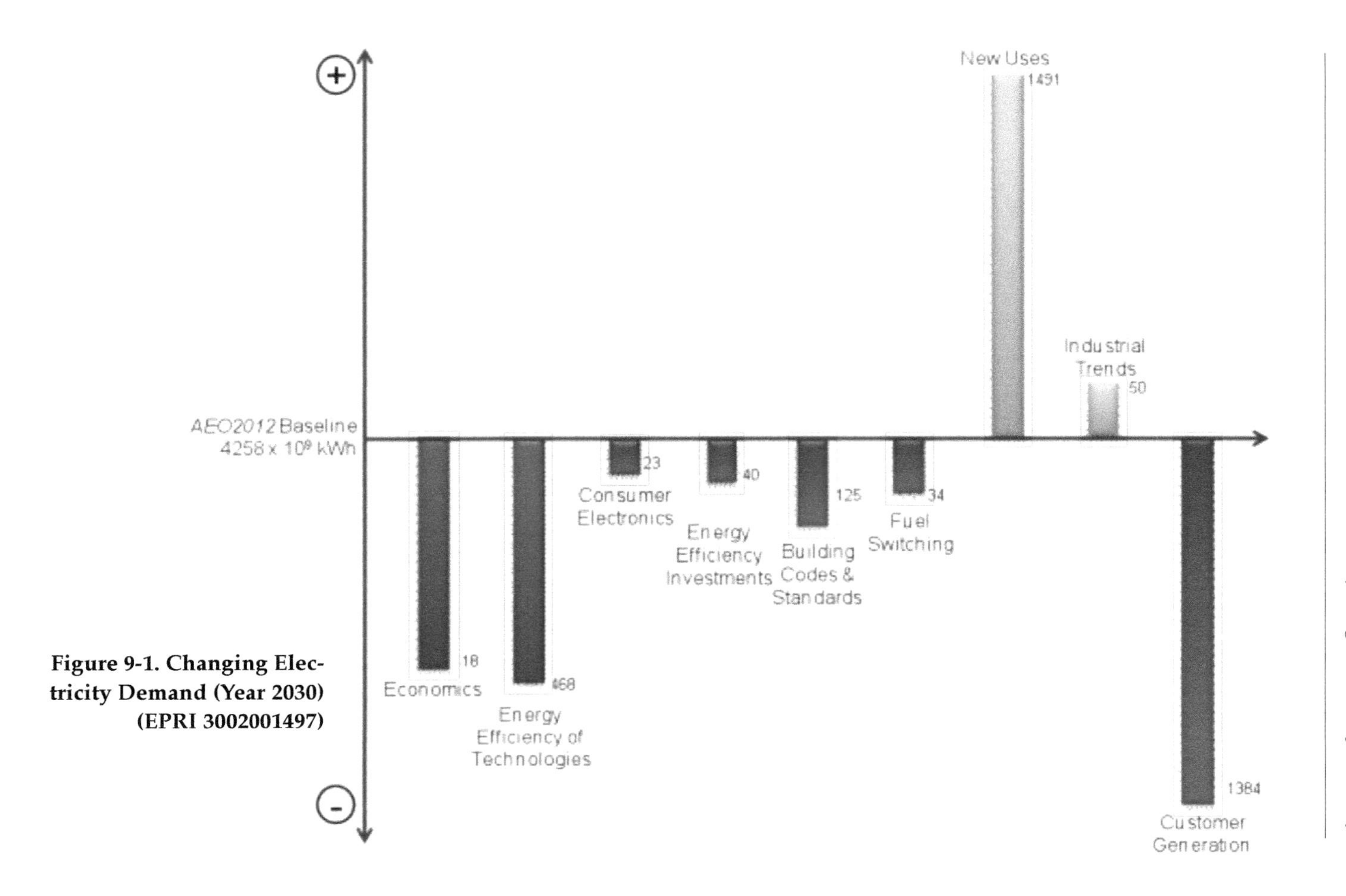

Figure 9-1. Changing Electricity Demand (Year 2030) (EPRI 3002001497)

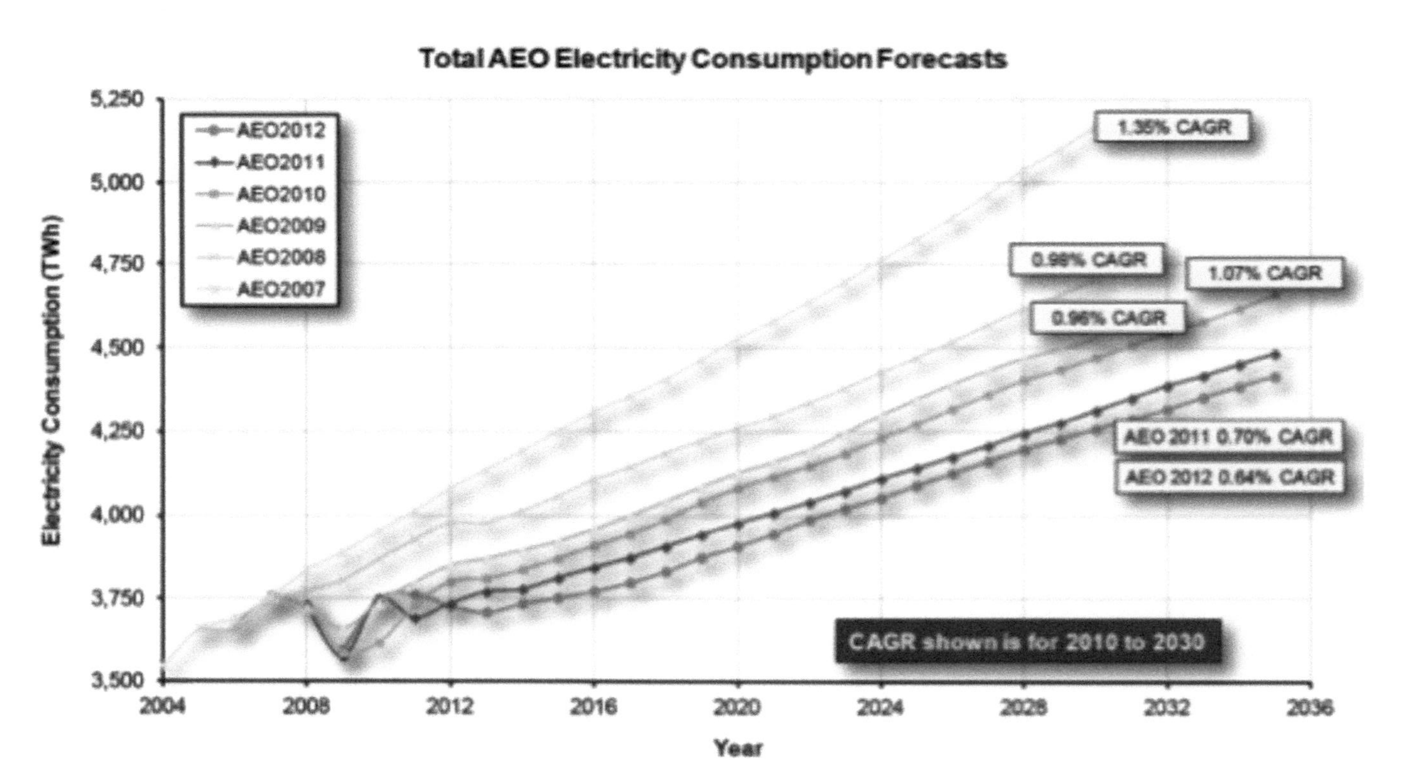

Figure 9-2. A Continuing Decline in the Projection for Future Electricity Demand (AEO 2012)

sale consumers, such as large industrials, are already reacting. In addition, housing trends are tending toward smaller units with fewer occupants. Increasing digital and entertainment purchases are propping up kWh consumption. Investments in the power system over the next 20 years will increase retail prices from 8.4% to 12.8% for residential and commercial consumers and reduce demand by between 284 and 418 x 109 kWh in 2030. The greatest promise for opportunities to enhance value may be embodied in the following potential changes in efficiency.

End-use Efficiency Improvements

Efficiency improvement continues in major end uses, and is driven by technology and energy efficiency program investments (see Figure 9-5).

Figure 9-3 illustrates some key technology trends not fully included in AEO2012.

Figure 9-4. Illustrates the potential technology diffusion which these technologies may enjoy.

Utilities and other energy service providers could engage in this evolving marketplace for energy efficiency by providing energy-efficiency related products and services.

Energy services companies can leverage the increasing funding available for energy-efficiency programs (see Figure 9-6)

Several wholesale markets now incorporate demand "resources" (energy efficiency and demand response) in their resource plans. While these are predominantly peak clipping activities, they have an impact on overall demand. Wholesale market programs can have an impact on energy demand potentially, reaching approximately 39 x 109 kWh in the year 2030.

Further evidence can be seen in reviewing the enormous impact which appliance efficiency standards have had on energy use. Twenty states have energy-efficiency resource standards (see Figure 9-7).

If the 28 states that do not have energy-efficiency codes or standards adopt them, a reduction of 125 x 109 kWh in 2030 is possible.

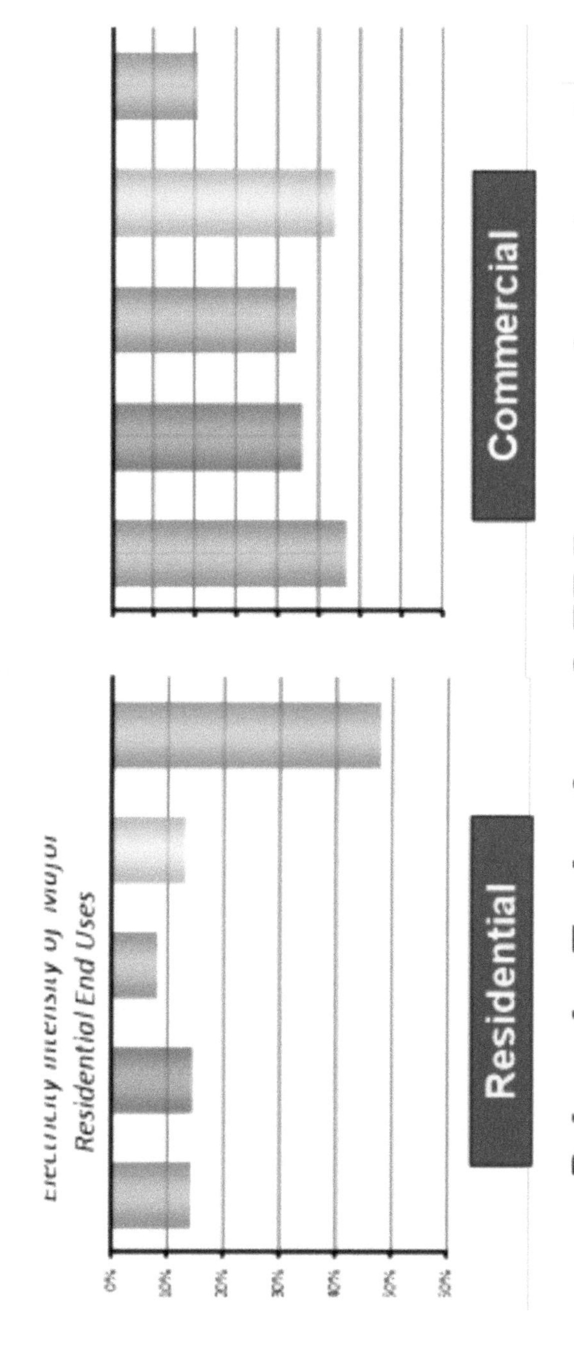

Figure 9-3. 2012 Annual Energy Outlook (AEO) 2030 Prediction (EIA)

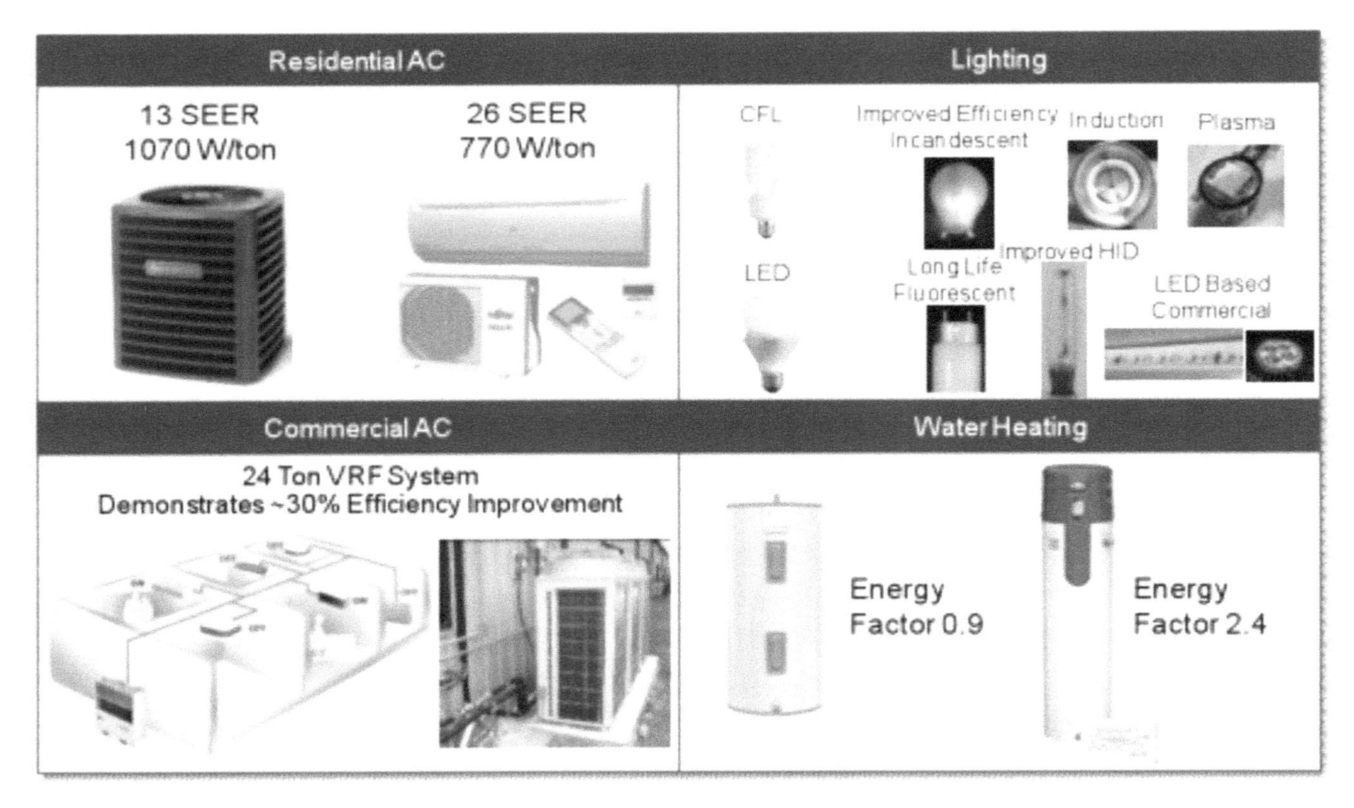

Figure 9-4. Technology Diffusion—Next Decade

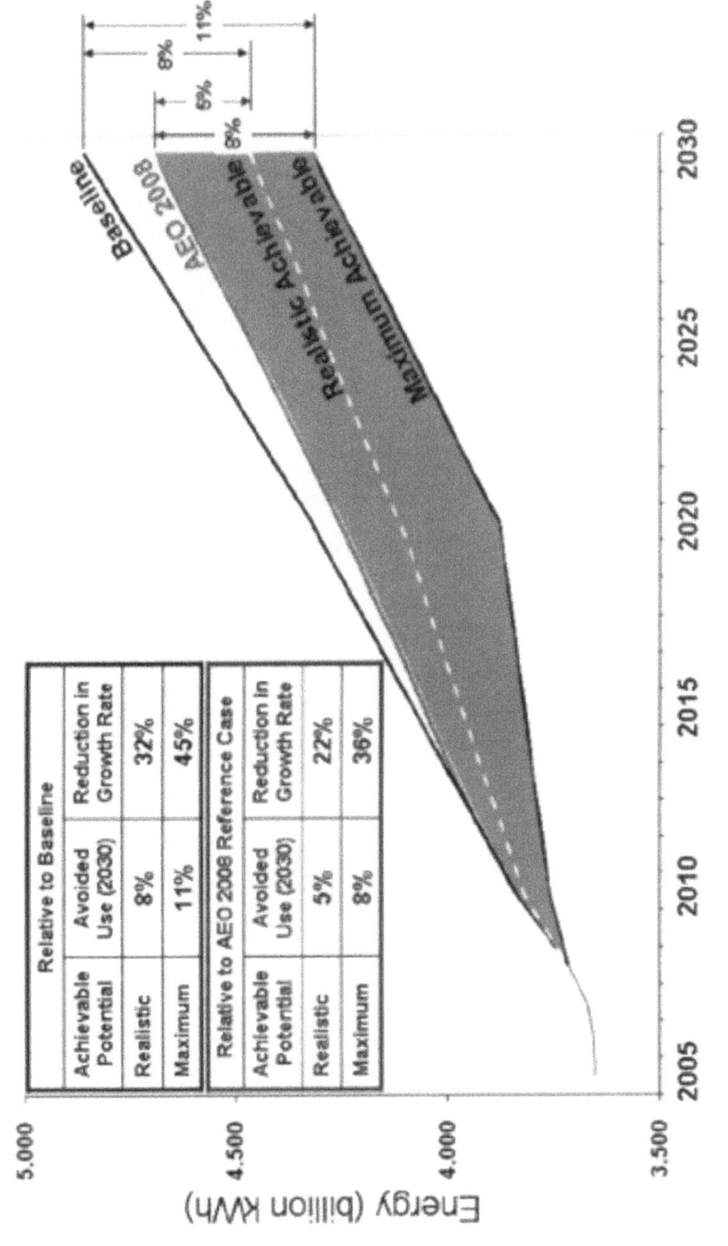

Relative to Baseline		
Achievable Potential	Avoided Use (2030)	Reduction in Growth Rate
Realistic	8%	32%
Maximum	11%	45%

Relative to AEO 2008 Reference Case		
Achievable Potential	Avoided Use (2030)	Reduction in Growth Rate
Realistic	5%	22%
Maximum	8%	36%

Figure 9-5. U.S. Energy Efficiency Achievable Potential (EPRI 3002001497)

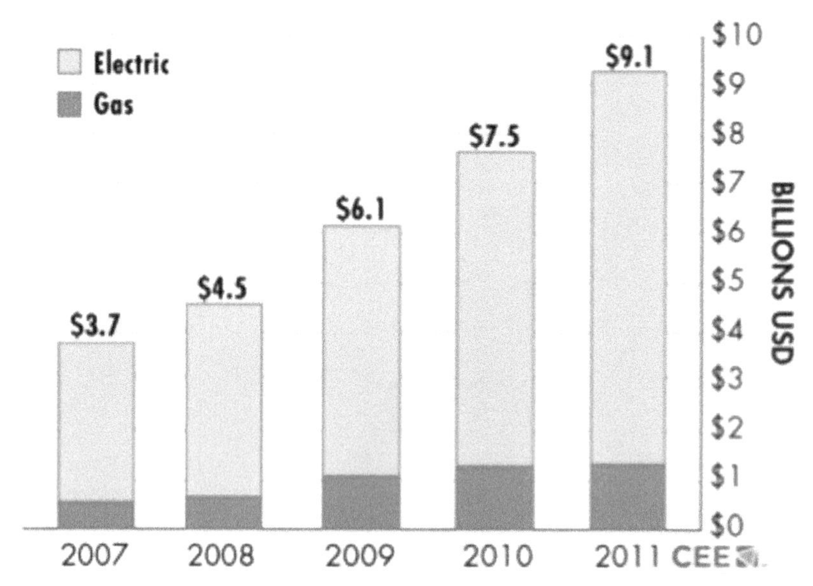

Figure 9-6. U.S. and Canadian Efficiency Program Budgets 2007-2011 (Consortium for Energy Efficiency (CEE)

How Can We Make Electricity More Valuable?

Electricity can be made more valuable by reducing the cost of electricity services through reductions in the price of electricity or increases in the value of the services it provides (enhance the service itself or improve the effectiveness or efficiency of end uses.) Electricity will also become more valuable by reducing the environmental impact of electricity production, delivery and use, e.g., improve end uses, and increase renewable and low-carbon generation. Alternatively, the value of electricity can be increased by increasing its reliability and/or increasing power quality by providing digital-grade electricity.

The value can also be enhanced through electrification which offers a net reduction in energy use and overall emissions or by deploying new uses which displace inefficient and ineffective fossil fuels. Value can also be enhanced by increasing the integration of information technology so as to constantly improve the electric purchase experience—enhance customer connectivity.

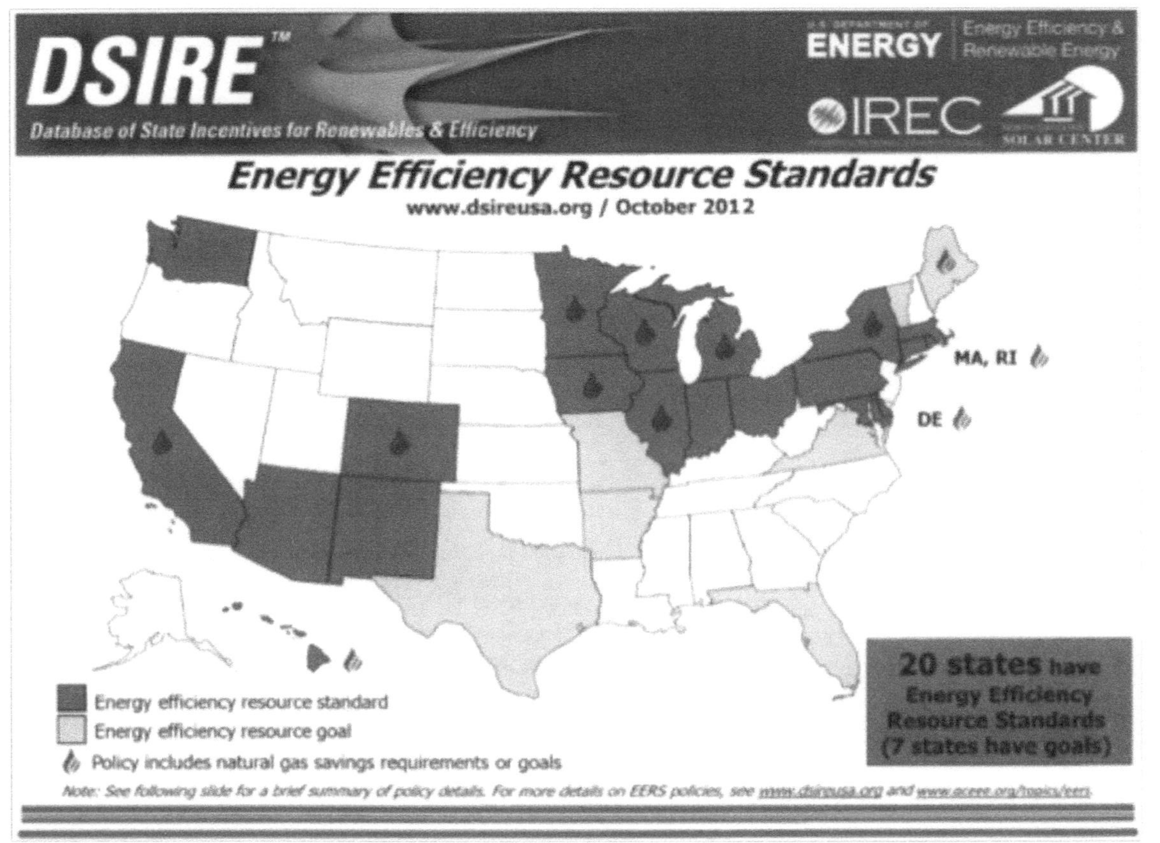

Figure 9-7. Energy-Efficiency Resource Standards (U.S. DOE)

IN SUMMARY, ELECTRICITY'S VALUE CAN BE ENHANCED BY THE FOLLOWING ACTIONS:

— Make existing uses of electricity as efficient as practical.

— Assure that new uses are as effective as practical.

— Develop new uses that displace the use of natural gas while reducing net energy consumption and emissions.

— Expand the use of renewable power generation throughout the power system.

— Effectively integrate large-scale renewable power generation with the power system. Includes developing balancing resources to manage intermittency such as storage and fast-acting demand response.

— Develop, demonstrate and deploy advanced cost-effective solar, wind and geothermal power generation technology.

— Enable two-way power flow on the grid while leveraging various combinations of central generation, microgrids and generation at customer locations.

— Evolve today's power system into plug-and-play power systems where all components ranging from end-use devices, distributed resources, T&D equipment, and power generation or storage can be "plugged in" virtually anywhere and be immediately inter-operative—analogous to the telephone and the internet.

In order to realize the electric future, two levels of functionality are needed. The first level is essentially that which has currently been defined as being associated with the Smart Grid. This includes the following attributes (EPRI Report to NIST, 2009):

- Enable Active Participation by Consumers
- Accommodate All Generation and Storage Options

- Enable New Products, Services, and Markets
- Provide Power Quality for the Digital Economy
- Optimize Asset Utilization and Operate Efficiently
- Anticipate and Respond to System Disturbances [Autonomously] (Self-heal)
- Operate Resiliently Against Attack and Natural Disaster

The second level of functionality requires the power system to become truly "plug-and-play." This includes the ability to electrically plug in any end-use electric appliance or device so that it is automatically "connected" from both a communications perspective and an energy perspective (i.e., energized) including devices which:

- Demand electricity
- Generate electricity
- Store electricity
- Are synergistically capable of demanding, generating and storing at various times
- Enable all major end-use devices and distributed energy resources to be connected from a communications perspective and are able to both receive signals and send information.
- What needs to be done to make this a reality?
- Standardize communications
- Develop the needed consumer portals
- Big data management and analysis
- System stability
- Manage balance of supply vs. demand

Chapter 10

Open Market Energy Consumer

In three historical "waves" during the last century, America sharply redefined its non-transportation energy supply system. Each "energy wave" has changed not only how we provide energy, but how we use it, with those forces working interactively. As these changes continue to evolve, they will impact the value of electricity, steadily increasing its importance to humankind.

Today, Americans spend nearly $1 trillion annually on non-transportation energy and the devices that consume it—only a third of which comprises traditional energy utilities' revenue. Through a confluence of emerging technologies, growing consumer awareness, and a changing regulatory framework, a fourth energy wave is developing—the Open Market Energy Consumer. This will change how consumers and suppliers will interact and speculates that dramatic results could emerge as this marketplace develops. No longer can electricity be considered simply as a refined energy form, but all of the attributes of electricity, including the technologies that are used to delivery, utilize, locally produce, store, purchase and administer it have become part of the energy marketplace. As such, the potential to enhance its value have expanded as well.

The U.S. economy depends heavily upon a safe, reliable, and economical energy supply. Americans enjoy the fruits of a system of energy production, distribution, and use that is without peer, and they use it heavily. Americans use energy for comfort and convenience; for production and distribution of goods and services; and for creation, processing and distribution of information. Energy costs in the U.S. are relatively low and stable, and energy supplies are admirably reliable. The U.S. energy supply system works well. Yet we are setting out to "fix it" by broadening the recognition of its attributes, even though it appears at first glance that it "ain't broke."

ENERGY WAVES AND THE OPEN MARKET CUSTOMER

The electric energy business is scarcely a century old. Born in an era of proud individualism, it went through successive phases of development triggered first by a changing political and regulatory framework, and then by unexpected external developments. With a bow to his seminal thinking, we have borrowed Alvin Toffler's concept of "waves" to define these phases. Like real waves, these four energy waves look alike superficially. However, each has defining and unique characteristics, and each obliterates what came before it.

The First Energy Wave

The **First Energy Wave** was created by the infant electric energy industry of 1880 to 1910. Utilities came about almost as an afterthought from the mind of inventor Thomas Edison. His main goal was to create a functional electric light bulb—a task that took him 8,000 trials of different filament materials before he succeeded in 1879. Three years later, he threw a ceremonial switch to inaugurate electric service from the Pearl Street Generating Station in New York City. The nascent electric utility industry fostered technological competition—alternating current versus direct current, ever-larger generators, and various prime energy sources—and competition for market share. Entrepreneurs in cities large and small jousted to provide bundled generation, distribution and transmission services, and seeking competitive advantage. Attributes of value during this wave have been well documented in previous chapters of this book and include some of the most basic advantages of electricity such as its ability to provide artificial illumination and motive power.

The Second Energy Wave

The **Second Energy Wave** brought order out of the chaos left on the beach as the first wave receded. Inevitably, the growth created pressure for combination of small electric supply and demand companies into ever-larger enterprises. With mergers came growing public concerns about potential abuses of market power. Already enamored of government regulation because of experience with railroads, Americans turned to regulation as a means to control the perceived excesses of the electric utility industry. Rising populist sentiment resulted in some electric utilities becoming a function of government.

The Second Energy Wave brought technological progress, consolidation, and extension of service to all but the most remote locales. Its utility regulation brought the concepts of "obligation to serve" and "return on capital" as the yardstick for setting rates—a powerful incentive for utility executives to maintain vertical integration and encourage consumption. This era of progress, stability and growth proceeded from 1910 to 1970, interrupted only by the financial setbacks of the Great Depression. Utilities went for years and even decades without raising rates. Utility stocks were considered safe harbor for widows and orphans. Nuclear energy offered the promise of "energy too cheap to meter." The nation's energy system appeared immune to serious disruption.

During the second wave, the value of electricity was expanded not only but making electricity available to nearly all humans in the western world, but through its evolving applications by leveraging all aspects of the electromagnetic spectrum. Principles of microwaves, radio frequencies, infrared and ultraviolet energy allowed enormous changes in industrial productivity and even home appliances.

The Third Energy Wave

The **Third Energy Wave** proved that appearances were deceiving. The formation of OPEC and subsequent oil embargoes triggered the infamous "energy crisis" in the U.S. and throughout the industrialized world. Raging inflation and 20% interest rates ensued. Growing environmental concern about air, water and soil pollution focused on electricity production as the root of much environmental evil. Nuclear power suffered the double blows of Three Mile Island and Chernobyl.

During this wave, the value of electricity increased both continued electrification or the displacement of inefficient fossil fueled applications by electrical applications. The result of this electrification was an enormous increase in the amount of Gross Domestic Product (GDP) produced per unit of energy input.

Policymakers reacted to these developments by imposing an even heavier regulatory hand upon the utility industry. President Jimmy Carter deemed energy conservation "the moral equivalent of war" and signed legislation that helped create two new industries—demand-side management and renewable energy. Air quality regulations mandated expensive emission controls. Safety enhancements drove the cost of nuclear power plants to high multiples of their originally estimated cost. Many states embraced processes of centralized planning (under the rubric of Integrated Resource Planning) for power plant siting and development.

Clearly, the shocks of OPEC, TMI, environment concerns, and inflation "broke" the Second Energy Wave system despite its seeming invulnerability. But the Third Energy Wave presented its own challenges. That is why we now are witnessing the first cresting of...

The Fourth Energy Wave

The **Fourth Energy Wave**—a re-envisioning of the energy industry characterized by reliance on market forces for pricing, resource planning and energy acquisition, and enhanced opportunity for innovative service and product design. In the fourth wave, "obligation to serve" becomes "privilege to serve." The fourth wave refocuses the industry from societal impacts to markets and customers. Dynamic market-responsive cultures replace lethargic central planning cultures. We have chosen to call this new environment the "Open Market Energy Consumer Industry." Its habitants might be called "Open Market Energy Consumers" and "Energy Entrepreneurs." The latter category encompasses *all suppliers* in the marketplace; we define them as *entrepreneurs* because, whether small or large, startup or established industry leader, they must *behave* as nimble, aggressive companies if they are to prosper in this newly competitive arena. These buyers and sellers will reinforce each other's goals, purposes, and initiatives to create a Fourth Energy Wave that will be different from its predecessors.

It is in this energy wave that all the attributes of the energy marketplace can be fully recognized and embraced, allowing an expanded view of the ultimate value of electricity.

To consider the implications of the Open Market Energy Customer Industry and the expanded view of the value of electricity, one must first understand its scope and size. Analysis reveals that the energy functions undertaken by traditional energy utilities represents only *one-third* of the total marketplace. The Open Market Energy Customer Industry encompasses all energy users and all organizations with whom energy users must interface, directly or indirectly, to obtain desired energy services. This includes:

- Providers of electric, natural gas and related raw energies to residential, commercial and industrial energy users.
- Providers of equipment, technology, products, services and systems needed to convert energy to meet the customers' needs including local generation and storage providers.
- Providers of design, construction, operation, maintenance and finance of such products and systems.
- Those who provide knowledge and information to the energy marketplace including providers of energy, energy technologies and services and consumers themselves.
- Federal-, state- and local-level regulators of the marketplace and its habitants.

The New Energy Marketplace

Figure 10-1 presents this new marketplace as nine individual business functions or market segments. One can think of the

Open Market Energy Consumer Industry as having three basic challenges.

- **Obtaining Commodity Energy:** Having inexpensive and reliable raw energy (primarily electricity and natural gas) delivered to industry, office or home (Market Segments 1-3).
- **Adding Value:** Buying, installing, operating and maintaining all the necessary apparatus to control the energy and convert it to the services desired (Market Segments 4-7).
- **Financing:** Financing and administrating the energy infrastructure and equipment in value-added category (Market Segments 8-9).

Table 10-1 lists an estimate of the distribution of annual expenditures and total investment by energy consumers. In this table, commodity energy is only a third of the total marketplace. Value-added services and financing them is two-thirds. Equally significantly, the American energy consumer has sunk huge capital investments in energy infrastructure and energy-using equipment.

While the first three line items in Table 10-1 are reasonably reliable estimates, lines 4 through 9 are less precise. Lines 4 through 9 are indicative of orders of magnitude rather than as precise market measurements. Nevertheless, they reveal important aspects of the total Open Market Energy Consumer Industry.

Each segment of the Open Market Energy Consumer Industry comprises a myriad of companies and organizations. They range from small niche players that serve a specific need in a local geographical area to global giants with multiple product and service offerings. Each segment has its own inputs and outputs, and its own definition of success. Companies will succeed in this marketplace to the extent that they deal effectively with perceived and actual customer needs. To a considerable degree, however, consumers in an emerging market don't know what they need until they are shown it.

Today's energy consumer is largely indifferent to the existing energy marketplace because he has few options. Experiences of

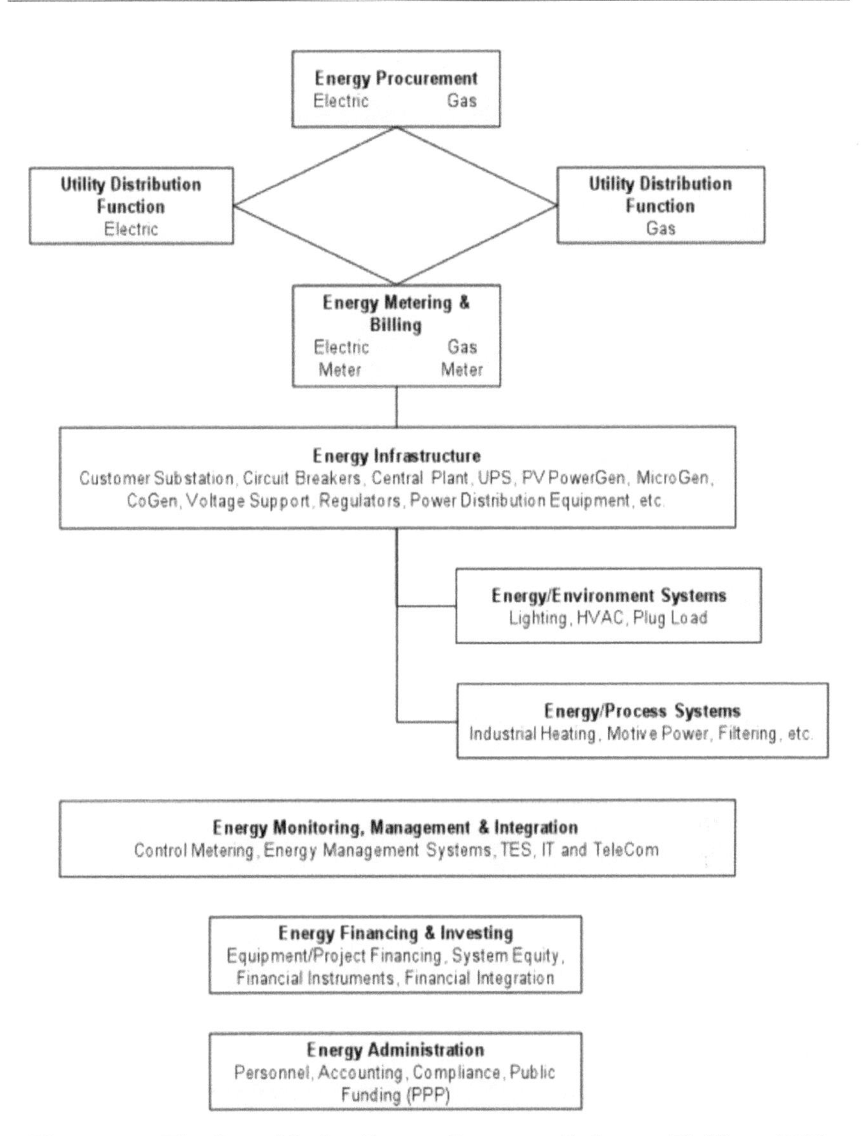

Figure 10-1. The Open Market Energy Consumer Industry (Gellings 1997)

other industries show that as new products and services emerge in a competitive environment, they can create significant new demand and revenue streams. Before the breakup of AT&T, who would have thought about having voice mail on their residential phones? Or call waiting, forwarding, screening, speed calling,

Table 10-1. Business Functions in the Open Market Energy Consumer Industry (Gellings 1997)

These percentages are beginning to change in certain geographic areas. As consumers become more like prosumers, by buying, leasing or otherwise engaging in the production and storage of energy, more of the energy expense will go toward the expanded use of technology, its financing and administration.

Business Function	Expense % of Total
Energy Procurement	14%
Utility Distribution Company	20%
Energy Metering and Billing	1%
Energy Infrastructure	17%
Energy/Environment Systems	13%
Energy/Process Systems	3%
Energy Monitoring/Management/Integration	1%
Energy Financing and Investing	29%
Energy Administration	4%

repeat dialing, and priority ringing? A cellular phone in their car? Voice mail for the cellular phone? People are spending much more on communications services than ever before *because they are buying new and different services* that they never thought about until the services became available.

Issues Facing the Open Market Customer

As the Open Market Energy Consumer Industry becomes more mature, consumer choices will expand exponentially. Consciously, or unconsciously, each energy consumer must begin to redefine his energy needs. Residential consumers are likely to remain largely indifferent to their energy options, if their activity in the telecommunications marketplace is any indicator. Larger commercial and industrial users, especially those for whom energy is a large part of their variable costs, will pay much more attention, just as they do today. Whether consumers simply throw up their hands in disgust (as many have done with respect to long distance telephone service) or become active participants will depend significantly upon the ability of marketplace participants to communicate the benefits of particular service and products.

Consider the range of issues confronting the Open Market Energy Consumer Industry:

- **Energy Purchase:** Price. Marketer reliability. Contract Terms.
- **Energy Delivery:** Utility distribution company reliability and responsiveness in times of emergency. Technical interface between user's system and the distribution company. Distribution company engagement in providing and servicing distributed energy resources.
- **Metering:** Accuracy. Timeliness of reading. Information value of reports and bills.
- **On-site Energy Generation, Control and Distribution:** Local power generation and storage. Maintenance. Emergency generation capabilities. Fuel flexibility. Building energy management services, building commissioning. Infrastructure system renewal plans. Power quality.
- **Energy Transformation:** Lighting, heating and cooling systems adequacy, efficiency, and cost effectiveness. Maintenance needs and plans.
- **Industrial Process Use:** Technologies to improve productivity, product quality and environmental regulation compliance.
- **Monitoring and Measurement:** Data adequacy, appropriateness and value. Invest versus buy tradeoffs. Efficiency enhancement versus infrastructure expansion tradeoffs. New versus old energy technology tradeoffs. Telecommunication and computer technology integration with energy systems.
- **Financing and Investing:** Energy cost hedging options. Lease versus own. In-house versus outsourcing. Finance versus invest. Capitalization versus expensing. Financial

services bundling. Shared savings versus performance contracts.

- **Administration and Overhead:** Personnel competency and training. Staff size. Accounts receivable and payable management. Regulatory Compliance awareness. Incentive program awareness. Energy options awareness.

Customer Expectations

To paraphrase Abraham Lincoln, most consumers will ignore some of these issues all of the time, and all of these issues some of the time, but they *won't* ignore all of these issues all of the time. It is when they focus upon these topics and their ramifications that opportunities will arrive in the Open Market Energy Consumer Industry.

Studies show that most consumers care about energy service only when it disappears (a power interruption) or when they have to pay for it (when they receive their energy bill). Only large commercial and industrial energy users, or sophisticated small business users, are actively involved in one or more of these issues daily. However, this is changing dramatically as the better educated, technically adept consumers are equally interested in their domestic energy use.

Indeed, customers' awareness and attitudes are changing quickly. With the deregulation of other industries, consumers have quickly become knowledgeable and discriminating customers of the goods and services presented to them. From the telecommunication industry to the airline industry, consumers have proven to be aware and responsive to the latest in technology innovation and marketing programs.

WHAT WILL BE THE KILLER "APPS" OF THE EMERGING OPEN MARKET ENERGY CONSUMER INDUSTRY?

The short answer is that no one really knows. A relevant consideration is the sectors of the marketplace that are likely to contribute to significant reshaping of the industry in the Fourth Energy Wave. Here, as with identification of potential Killer Energy Applications or Apps, one makes forecasts only with trepidation. Nevertheless, the author here ventures such predictions. Here is a Letterman-like list of "Ten Predictions about the Fourth Energy Wave":

- **Technologies that convert energy to services and those that generate and store electricity locally will go through a resurgence of research, development, demonstration and successful commercialization.** Energy/Environmental systems and Energy/Process systems have historically enjoyed a high growth rate of development and implementation. Lower prices will significantly improve the cost effectiveness of many of the energy process technologies compared with non-energy industrial processes (mechanical, chemical, etc.). As sales of these technologies increase, state and federal regulators may be more inclined to mandate technologies for environmental improvements—further reducing equipment prices.
- **Energy-efficient lighting and heating/ventilating/air-conditioning (HVAC) technologies have benefitted from utility rebates for the last several decades.** More recently, distributed generation and storage are enjoying substantial incentives as well. While such rebates and incentives will likely change over time, other forces will foster new development in these technologies. These forces could be (1) a renewed energy services company industry spurred by the retail affiliates of energy companies; (2) the "public-purpose program" that are developing in more and more states as regulators cling to the Energy-Efficiency and Renewable Energy ethic; and (3) continued interest in federal and state appliance and building standards, incentives for renewables and renewable portfolio standards.
- **The *quality* of energy could become a significant price and service differentiator.** In the DSM and load management days of Energy Wave Three, some electric utilities were sur-

prised by the robust customer response to interruptible tariff offerings. At the same time, other customers were paying for additional facilities to assure uninterruptible, digital grade service. This demonstrates potential demand for a wide variety of "power quality" and a ready market for interruptible power, firm power, standby generation, co-generation, uninterruptible power systems, clean power technologies and similar offerings. The business segments of energy procurement, utility distribution, energy infrastructure, and energy system integration could all foster power quality services. Appropriate integration of each of these options could well become a significant service in the Open Market Energy Consumer Industry.

- **The open energy marketplace will [continue to?] unite with the information management and telecommunications industries to produce dynamic new products and services.** One of the more exciting opportunities in the Open Market Energy Consumer Industry will be in the monitoring, management and integration businesses area. Data management computer control systems and telecommunication medium (microwave, fiber optic, power line modulation, etc.) systems will provide almost unlimited opportunities in residential, commercial and industrial marketplaces alike. Customers' multi-site locations across the U.S. will have instant monitoring, management, and integration capabilities that will allow them cost and service enhancements unfathomable now. Residential appliances could well have internet-type addresses that could transmit data over house wiring infrastructure to smart meters. This data would produce information on appliance usage and maintenance and replacement opportunities. Knowing when and for how long refrigerator doors are open could infer food use patterns and help supermarket chains target market. It is not hard to imagine that such information would have more value (to equipment manufacturers, distributors, or marketers) than the actual energy used by the appliance!

- **The energy metering business area will continue to experience dynamic growth and renewal.** One of the most important and growing business areas will be the energy metering function. The right to provide this function, long known as the "cash register" of the energy supply industry, is being heavily contested in the utility industry restructuring process. Its position as the primary interface between the upstream commodity supply functions and the downstream energy service functions is critical. Obviously the establishment of hourly pricing is going to create (has created) a demand for the replacement of millions of energy meters... but that is only the beginning. One can speculate about the impact of aggregation of multi-site customers, electronic billing, and smart meters that choose the best of several pre-selected supply options. It is difficult to believe that with all the advances in technology each American home will continue to have a separate meter for electricity, gas and water.
- **Consumers will increasingly install local generation and storage utilizing distributed photovoltaics, micro generation, combined heat and power systems, and/or batteries and thermal energy storage.**
- **Consumers will devote greater attention to the management of their energy infrastructure.** Like metering, this functional business area defines a critical juncture in the integration of all energy and energy service solutions. Sound management practices will give large energy users greater flexibility to operate upstream (energy procurement) and downstream (energy conversion) as their local conditions may warrant. On-site generation will experience growth because:
  - Consumers requiring local on-site reliability.
  - Consumers wanting the ability to sell into the energy market.
  - New technology development (efficient small generation systems and new technologies, e.g., fuel cells). The Open Market Energy Consumer Industry participants, especial-

ly those with utility background, will have opportunities to provide outsourcing services for design, construction, operations and management of customer-owned energy infrastructure.

- **Suppliers will use innovative marketing strategies to sell into the open energy marketplace.** The Open Market Energy Consumer Industry will be the beneficiary (or victim) of some of the same innovative marketing strategies that other deregulation has produced (e.g., airlines and telecommunication). This may be particularly true in residential and small commercial and industrial markets where mass marketing will prevail. Consumers will be able to choose a wide variety of energy marketing options (e.g., "a free airline mile for each kWh used"). These marketing strategies will be combined with new technologies to create a "first-on-the-block" syndrome.
- **More energy consumers will take advantage of energy and energy service outsourcing.** Consumers will be presented with a greater variety of choices in all of the business areas. This will produce consumer frustration from lack of familiarity with energy and energy service issues. Except for the largest players, consumers may not want to take the time to fully comprehend the impacts of various options to their business and will request expert assistance. Residential and small commercial and small industrial businesses will especially be more inclined to look at outsourcing the energy and energy services function (most likely through affinity groups). This will lead to significant bidding and selection of energy partners with groups of small customers.
- **A plethora of financial offerings will creatively respond to financial and investment challenges.** The amount of funds expended and invested by the Open Market Energy Consumer Industry will be an incentive for the financial services business to develop new financial offerings for energy consumers. These will be designed especially to attract residen-

tial and small commercial and industrial energy consumers. Creative financing could involve packaging both energy procurement expenses with energy equipment capitalization. Financial arrangements could also involve new forms of packaged energy and energy service such as end-use metering.

- **Mergers, partnerships and collaborative of various product, service and information companies and organizations will become commonplace throughout the industry.** Many "natural" industry mergers are already beginning: electric/gas utility with electric/gas utility; utility with power marketer; and utility with a variety of energy service companies. Over time, this list will extend throughout the industry. Rather than focusing on mergers of old utilities, a new series of partnerships will include equipment manufacturers, energy suppliers, financial service businesses, infrastructure engineering and construction companies, and system integrators. Project-specific, customer-specific, and technology-specific collaborations will become commonplace as players experience the strength of a dynamic list of partnerships. It is likely that these partnerships will extend to other industries that have synergistic advantages (water, wastewater and telecommunications, for example). The second and third wave employee cultures of entitlement and lifetime employment will change to a culture of a dynamic and mobile workforce. This will build natural relationships between market players which will foster still greater collaboration and partnerships.
- **Opening the upstream (commodity end) to the Open Market Energy Consumer Industry will open the downstream (value-added) end of the marketplace.** After energy customers exercise their freedom of choice in energy procurement, their remaining options to reduce energy costs and improve service will involve energy use. They will no longer be able to blame any high energy service costs on an inefficient local energy utility and its energy tariffs. As costs on the frontend

come down, costs on the backend proportionally will be more significant. A host of private and public agents will emerge to educate users about methods to control energy consumption. Potential profits will spur product offerings for design, construction, operations, maintenance and financing services for downstream equipment systems.

CONCLUSION: THE RATIONALE FOR AN OPEN ENERGY MARKETPLACE

U.S. economic success depends—along with many other factors—upon a reliable, flexible and low-cost energy supply. Although U.S. energy prices are relatively low and stable, and our electricity supply reliability is the best in the world, we cannot stand still. Global competitors constantly challenge U.S. businesses to realize the Olympic motto: *"Citius, Altius, Fortius"* (Swifter, Higher, and Stronger).

Within that context, there is a framework of interactive forces that together are pushing our traditional energy and energy-related industries into a new competitive arena for the Open Market Energy Consumer Industry:

The relationship between these five drivers is complex and interactive. Figure 10-2 suggests the paths of feedback that can occur. The industry structure is highly dynamic and volatile with respect to potential relationships between players.

While this suggests the direction in which this new marketplace may develop, it is likely that in many respects it will ultimately mature as something quite unlike even the most radical scenarios described today.

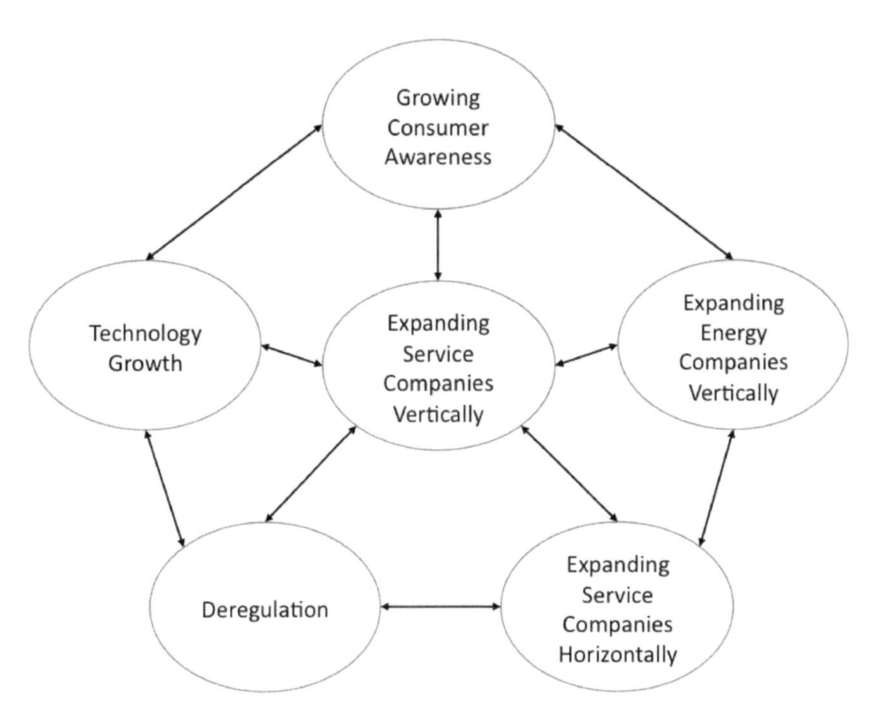

Figure 10-2. Dynamic Interactions in the Open Energy Marketplace (Gellings 1997)

Chapter 11

Summary of the Value of Electricity

It is difficult to organize all of the material contained in the previous chapters into one set of estimates. It is sufficient to say that electricity has substantial value. Tables 11-1 and 11-2 summarize the key attributes of the value.

VALUE BASED APPLICATIONS

Table 11-1 highlights the various estimates which had been assembled based on various applications of electricity. Among the most startling estimates come from these applications which directly impact longevity. The economic benefits from longevity are based on a 2004 study by the U.S. Environmental Protection Administration's (EPA) Office of Air and Radiation which estimated a value of $5.5 million for a statistical human life.

Table 11-1. Value of electricity based on various applications of electricity

Application	Value
Value in Saving Lives (20 years)	$248 trillion
Value in Self-Driving Cars	$4660 B/year
Value in Document Transfer	$70 B/year
Value in Reducing DUI Deaths	$23.4.B/year
Value in Rescue Communications	$20.4 B/year
Value in Smoke Detectors	$15.6 B/year
Value in Displacing Gasoline	$388 million/year

VALUE BASED ON SURVEYS AND ANALYSES

Table 11-2 summarizes the results of various surveys and empirical analyses on the impact of outages expressed in dollars per kilowatt hour. These estimates, like others, vary widely ranging from zero to several thousand dollars per kilowatt hour. Typically the lowest costs are for residential consumers experiencing short outages. There are little data available from outages exceeding 8 hours.

Table 11-2. Summary of Results from Surveys and Studies

Attribute / Study	$/Kwh (unless otherwise stated)
The value of electricity in powering air conditioning	0.54
Human Power Equivalent	.0972
Household Productivity	3.34
Life in Expectancy	52.80
Value in Freeze Concentrating Food	.068
Value in Displacing Fossil Fuels for Heating	0.68
Value in Self-Driving Cars	$4660 B/year
GDP Driven by Electricity	4,626
LBNL Voll Studies	.092 – 11.29
Customer Damage	2.40 – 365.50
UK Studies on Interruptions	.19 – 116.00
Finish Studies of Customer Losses from Outages	3.12 – 8.88
DOE ICE Calculator	.99 - 2665
ERCOT Value of Lost Load Study	0 – 539.07
London Economics Based on Value	82.39 – 170.13
Ontario Hydro Outage Costs	.04 – 150.56
DFI Study	.30 – 16.34
EPRI Study – hypothetical	4.08 – 5.03
EPRI Study based on Blackouts	.17 – 7.88
Analysis by Sullivan	1.00 – 2641.00
Meta Study (Literature)	.15 – 39.40

VALUE BASED ON WILLINGNESS TO PAY

Table 11-3 lists the result of a group of studies and surveys which estimate the willingness of consumers to pay additional fees for reducing or eliminating outages.

Table 11-3. Willingness to Pay (WTP) estimates

WTP Estimated	$ / Kwh
Asian Development Bank	26.40
NYC 1977 Blackout	16.15
Sweden	1.26 – 19.16
Based on TVA Survey	0.4 - 2.46

VALUE BASED ON IMPACT OF OUTAGES

Table 11-4 lists a variety of overall estimates of outages for both individual events and overall poor power quality and reliability.

Table 11-4. Results of a variety of overall estimates of outages

Estimates	Value
Storms – loss of more than 1 million customers up to 45 days in duration. Includes outages & Poor PQ from Surveys	$150 - $237 B/year
Galvin Estimates	$150 B/year
Large Industrial Customers	$142 M/day
Interruptions (LBNL)	$98 B/year
Impact from weather-related outages	$18 - $75 B/year
Regional Outage	$16 B/outage
Data Center Outages – Per Center/Outage	$1.7M
London Economics Study/Residential Customers Outage	$50/customer/outage

References

"Utility Marketing Strategies: Competition and the Economy," C.W. Gellings, 1994, Fairmont Press, Lilburn, GA, ISBN 0-88173-156-0.

"Effective Power Marketing," C.W. Gellings, 1998, PennWell Publishing Co., Tulsa, OK, ISBN 0-87814-646-6.

Wikibooks.org/wiki/Principles_of_Economics, 2/8/2012.

"Value of Service Reliability to Consumers," Seminar proceedings prepared by Criterion, Inc., EA-4494, March 1986.

"Value-Based Utility Planning: Scoping Study," Final report for EPRI RP-2381, prepared by Levy Associates and Meta Systems, EM-4839, December 1985.

"TAG™ Technical Assessment Guide—Volume 4: Fundamentals and Methods, End Use," C.W. Gellings, P. Hansen and S.M. Haas, EPRI August 1997, Report P-4463-SR, Volume 4.

"The Cost of Power Disturbances to Industrial & Digital Economy Companies," D. Lineweber and S. McNulty, 2001, EPRI Report 1006274.

http://en.wikipedia.org/wiki/The_Progress_of_Railroading.

Malthus, T.R., 1998, "An Essay on the Principle of Population," Chapter VII, Oxford World's Classics reprint.

"Humanity Unbound—How Fossil Fuels Saved Humanity from Nature and Nature from Humanity," I. Goklany, Policy Analysis, CATO Institute, Washington DC, December 20, 2012.

"Adapting to Climate Change: The Remarkable Decline I the U.S. Temperature-Mortality Relationship Over the 20th Century," A. Barraeca, et al., University of California Center for Energy and Environmental Economics (UCE3), http://www.uce3.berkeley.edu.

"What would happen if there was no electricity?," http://wiki.answers.com, May 20, 2013.

"The Instrumental Value of Technology: Technology and Christian Values," http://web.engr.oregonstate.edu.

"2001 Annual Energy Outlook (AEO)," U.S. Department of Energy, Energy Information Administration, April 2011.

"How does the body make electricity—and how does it use it?" http://science.howstuffworks.com.

"How does electric help save lives in hospitals?" http://boronj.lamascientists.org.uk.

"Does a Fear of Electricity Exist?" EZine Articles, http://ezinearticles.com.

"Electrophobia: Do you have a fear of electricity?" http://common-phobias.com.

"Fearing Electricity: Overhead Wire Panic in New York City," IEEE Technology and Society Magazine, fall 1995, The Institute of Electrical and Electronics Engineers.

David E. Nye, Electrifying America: Social Meanings of a New Technology, 1880-1940, Cambridge, MA: M.I.T. Press, 1990.

New York Times, September 27, 1889.

New York Times, October 15, 1889.

"Measuring the Benefits of Electricity, Energy for Development and Poverty Reduction," http://www.energyfordevelopment.com (EgyDev), February 8, 2010.

"Dark Light: Electricity and Anxiety From the Telegraph to the X-Ray," L. Simon, Harcourt Books, 2004. ISBN 0-15-100586-9.

"Empires of Light: Edison, Tesla, Westinghouse and the Race to Electrify the World," J. Jones, Random House Trade Paperbacks, New York, 2004.

War of Currents, Wikipedia, 2013.

"Executioner's Current—Thomas Edison, George Westinghouse and the Invention of the Electric Chair," R. Moran, Vintage Books, A Division of Random House, Inc., New York, 2002.

U.S. Patent Office, Patent Number 587,649.

"Edison Predicted It," New York Evening Sun, October 16, 1889.

A.P. Southwick to Thomas Edison, November 8, 1887, Edison Archives.

Harold P. Brown, "Death in the Wires," New York Evening Post, June 5, 1888, page 7.

"Lines Down—How We Pay, Use, Value rid Electricity Amid the Storm," S.A. Mitnick, Franklin Square Publishing, 2013.

"Famous Music Radio WABC Stories," WABC Radio 77, www.musicradio77.com/stories."Study: Proliferation of Home Air Conditioners Saves Lives," J. Ellperin, Washington Post, December 23, 2012.

Bureau of Labor Statistics, Consumer Expenditure Survey, www.census.gov. "The Connected Journey—Mobility at 40," R. Qubein, Business Traveler, July/August 2013.

Utility Marketing Strategies—Competition and the Economy, C. Gellings, Fairmont Press, 1994.

"Lamp Light on the Wabash," Y. Ksander, Moment of Indiana History, http://indianapublicmedia.org, July 7, 2008."Friday Night Lights," Back Story Radio, National Public Radio, July 2013.

"The Emerging Open Market Energy Customer: Market-Smart Consumers, New Suppliers, and New Products Will Combine to Shape the 'Fourth Energy Wave'," C. Gellings, EPRI and K. Gudger, Ken Gudger & Associates, 4th Annual Utility Strategic Marketing Conference, Orlando, FL, April 17-18, 1997.

"Utility Marketing Strategies: Competition and the Economy," C.W. Gellings, Fairmont Press, Lilburn, GA, 1994.

"Entergy Hurricane Preparedness," presentation by D. Hintz to EPRI Research Advisory Committee, April 18, 2006.

"TAG™ Technical Assessment Guide—Volume 4: Fundamentals and Methods, End Use," EPRI, Palo Alto, CA: August 1987. P-4463-SR.

"After the Disaster: Utility Restoration Cost Recovery," Edison Electric Institute, February 2005.

"Improving Electric Grid Reliability and Resilience: Lessons Learned from Superstorm Sandy and Other Event," Workshop Summary and Key Recommendations, The GridWise Alliance, June 2013.

"CLASSIFY-Profiles, Volume 1: Residential Customer Needs and Energy Decision Making," EPRI, Palo Alto, CA: December 1994. TR-104567-V1.

"The Program on Technology Innovation: Tracking the Demand for Electricity from Grid Services," EPRI, Palo Alto, CA: May 2013. 3002001497.

M. Sullivan and D. Keane, "Outage Cost Estimation Guidebook," EPRI, Palo Alto, CA: 2012. 106082.

"Incentive Regulation for Grid Reliability Helps Close Investment Gap," P. Centolella, NEMA Electroindustry, November 2012

"Microgrids: A Primer," A document prepared for EPRI member review and input, by H. Kamath and T. Key, EPRI, Palo Alto, CA: September 6, 2013

http://www.huffingtonpost.com/2012/10/30/hurricane-sandy-power-outage-map-infographic_n_2044411.html, Huffington Post, 10 October 2012. [Online]

J. Fahey, http://bigstory.ap.org/article/power-outage-time-after-sandy-not-extraordinary, 16 November 2012. [Online].

A. Bredenberg, http://news.thomasnet.com/green_clean/2012/11/12/could-decentralized-microgrids-solve-the-extreme-weather-outage-problem/, ThomasNet, 12 November 2012. [Online].

M.L. Wald, http://green.blogs.nytimes.com/2013/11/05/how-n-y-u-stayed-partly-warm-and-lighted/?_r=0, The *New York Times*, 5 November 2012. [Online].

W. Pentland, http://www.forbes.com/sites/williampentland/2012/10/31/where-the-lights-stayed-on-during-hurricane-sandy/, Forbes, 31 October 2012. [Onlint].

J. Tripolitis, T. Short and M. Olearczyk, "EPRI Distribution Grid

Resiliency Initiative," 2013-2014.

M. McGranaghan, M. Olearczyk and C. Gellings, "Enhancing Distribution Resiliency—Opportunities for Applying Innovative Technologies," EPRI, Palo Alto, CA: 2013.

"New Business Models for the Distribution Edge: The transition from value chain to value constellation," Rocky Mountain Institute, Boulder, CO, April 2013.

T.O. Paul, "State and Federal," U.S. Department of Energy, 07 November

2012. [Online].

Campbell, Richard J. "Weather-Related Power Outages and Electric System Resiliency," Congressional Research Service, August 28, 2012. Web site.

Sullivan, et al., "Estimated Value of Service Reliability for Electric Utility Customers in the United States," Ernest Orlando Lawrence Berkeley National Laboratory, June 2009.

U.S. Department of Energy, Office of Electricity Delivery and Energy Reliability, "Electric Disturbance Events (OE-417): Annual Summaries." Web site.

"Economic Benefits of Increasing Electric Grid Resilience to Weather Outages," Executive Office of the President, August 2013.

"Hurricane Sandy—Rebuilding Strategies: Stronger Communities, A Resilient Region," Hurricane Sandy Rebuilding Task Force, U.S. Department of Housing and Urban Development (USDHUA), 2013.

"Results and Findings from the ARRA-Funded Smart Grid Projects," presentation by J. Paladino, U.S. Department of Energy, EPRI Sector Council Meeting, Baltimore, MD, September 11, 2013.

"Capacity Markets in PJM," J. Bowring, *Economics of Energy & Environmental Policy*, Vol. 2, No. 2, IAEE, 2013.

"Capacity Market Fundamentals," P. Crampton, et al., *Economics of Energy & Environmental Policy*, Vol. 2, No. 2, IAEE, 2013.

Department of Energy, "Comparing the Impacts of Northeast Hurricanes on Energy Infrastructure," 04/2013, http://www.energy.gov/sites/prod/files/2013/04/f0/Northeast%20Storm%20Comparison_FINAL_041513c.pdf.

National Hurricane Center, "Tropical Cyclone Report: Hurricane Sandy," 02/12/2013, http://www.nhc.noaa.gov/data/tcr/AL182012_Sandy.pdf.

Analysis of Business Customers' Willingness-to-pay for Power System

Enhancements, EPRI, Palo Alto, CA: 2004. 1011363.

Understanding the Cost of Power Interruptions to U.S. Electricity Consumers, K. Hamachi LaCommare and J.H. Eto, Ernest Orlando Lawrence

Berkeley National Laboratory, University of California Berkeley, September 2004.

The Economic Costs of Unsupplied Electricity: Evidence from Backup Generation Among African Firms, M. Oseni & M. Pollitt, University of Cambridge, Energy Policy Research Group, November 2013,.

The Economic Cost of the Blackout: An Issue Paper on the Northeastern Blackout, August 14, 2003, ICF Consulting, Fairfax, VA, 2003.

"Southwest Power Outage Economic Cost Put at $100M," D. Jergler, *Insurance Journal*, September 13, 2011.

"The Price of Failure: Data-Center Power Outages Cost Sears $2.2M in Profit," J. Pletz, Chicago Business, June 4, 2013.

The Electric Power System is Unreliable, Galvin Electricity Initiative, www.galvinpower.org, November 28, 2013.

"Counting the Cost of Power Cuts," C. Ross, Salisbury Journal (UK), November 27, 2013.

2013 Study on Data Center Outages, Ponemon Institute, September 2013.

The Economic Impacts of the August 2003 Blackout, The Electricity Consumers Resource Council (ELCON), February 9, 2004.

"How Technology Saves Lives, E. MacDonald," FOXBusiness, January 21, 2011.

BOP Spending on Information and Communications Technology, World Resources Institute, Washington DC.

"Google is Working on a Technology That, If Perfected, Would Save 1.2 Million Lives Per Year," N. Carlson, Business Insider, March 11, 2013.

All the Numbers, All the Facts on Mobile the Trillion-Dollar Industry: Why is Google Suggesting to Put Your Best People on Mobile, Consultant Value Added (CVA), March 7, 2011.

Smoke Detector Alarms Save Lives—Which One is Right for You?, C. Giaimo, www.simplisafer.com, August 21, 2013.

"An Electrifying Awakening: Electrical Stimulation of the Spinal Cord Could Let Paralyzed People Move Again," E. Waltz, IEEE Spectrum,

November 2013.

"Electricity in Gynecology," J.H. Aveling, *The British Medical Journal*, May 26, 1888.

"The Atom May Save Your Life," S.H. Spencer, *The Saturday Evening Post*, July 20, 1950.

"Foundation: Test a Heart, Save a Life," L. Winkley, *The San Diego Union-Tribune*, February 10, 2013.

"Domestic Technology: Labor-Saving or Enslaving," J. Wajcman, *Technologies and Values—Essential Readings*, Wiley-Blackwell, 2010.

The Effects of Rural Electrification on Employment: New Evidence from South Africa, T. Kinkelman, August 2010.

"How Do You Live Without Electricity," A. Evangelista, *Backwoods Home Magazine*, January / February 2002.

"1 Out of 5 Forced to Live Without Electricity," *What News Should Be*, www.whatnewsshouldbe.org, November 29, 2013.

The Energy Access Situation in Developing Countries: A Review Focusing on the Least Developed Countries and Sub-Saharan Africa, a joint report of the United Nations Development Program and the World Health Organization, http://content.undp.org/go/cms-service/stream/asset/?asset_id=2205620, November 2009.

"Living Without Electricity," High Lonesome Ranch, Inc. Countryside Magazine, 1995 and 1997.

Living Without Electricity, S. Scott and K. Pellman, Good Books, Intercourse, PA, 1999.

Electricity, R. Robinson, Black Cat, New York, 1971.

Impacts of Temperature Extremes, C.R. Adams, Cooperative Institute for Research in the Atmosphere, Colorado State University, 2014.

J. Constine & G. Ferenstein, TC News, September 18, 2013, www.techcrunch.com.

Consumers in the County: Technology and Social Change in Rural America, R. Kline, The Johns Hopkins University Press, Baltimore, MD, 2000.

How to Live Without Electricity—And Like It, A. Evangelista, Lehman's, Kidron, OH, 1997.

Environmental Assessment of Plug-In Hybrid Electric Vehicles, Volume 1: Nationwide Greenhouse Gas Emissions. EPRI, Palo Alto, CA: July 2007. 1015325.

The Potential to Reduce CO_2 Emissions by Expanding End-Use Applications of Electricity. EPRI, Palo Alto, CA: 2009. 1018906.

Energy Efficiency Planning Guidebook. EPRI, Palo Alto, CA: 2008. 1016273.

Electricity Reliability: Problems, Progress and Policy Solutions, G. Rouse and J. Kelly, Galvin Electricity Initiative, 2011.

The Human-Powered Home—Choosing Muscles over Motors, T. Dean, New Society Publishers, BC Canada, 2008.

The American System to Mass Production: 1800-1932, Hounshell, pps. 89-91.

Lehman's: Simple Products for a Simpler Life, Kidron, OH, Spring C1308C.

Historical Origins of Food Preservation, B. Nunmer, National Center for Home Food Preservation, May 2002.

How Much is Clean Water Worth?—A lot say researchers who are putting dollar values on wildlife and ecosystems—and proving that conservation pays, J. Morrison, National Wildlife Federation (NWF), NWF News-and-Magazines, 2/1/2005.

New York in 1872: A City filled with drunks, Ephemeral New York, Word Press, July 18, 2011

American Demographic History Charterbook: 1790 to 2010, Chapter 6. Households, Relationship to Householder, and Home Ownership, C. Gibson, October 10, 2012.

Health Care and Social Change in the United States: A Mixed System. A Mixed Blessing, by: B. Fetter

E. Meeker, "The Improving health of the United Sates: 1850-1915," Exploration in Economic History, 9 (1972)

New York in the 1870s: Ephemeral New York, Wordpress

James McCage's Lights and Shadown of New York Life, 1873

Historical Divorce Rate Statistics, A.M. Jones, Love to Know, http://divorce-lovetoknow.com

Estimating the Economically Optimal Reserve Margin in ERCOT, The Brattle Group, January 31, 2014

Willingness to Pay among Swedish Households to Avoid Power Outages, F. Carlsson and P. Martinsson, Working Papers in Economics #154, Gothenburg University, December 2004

Measuring Willingness to Pay for Electricity, P. Choynowski, ERD Technical Note Series N. 3, Asain Development Bank, July 2002

The Next Greatest Thing, Rural Electrification Administration (REA), Washington, DC, 1943

MBA in the Global Energy Industry: Operations Management—Electricity Distribution, J. Reneses & P. Frias, Univsersidad Pontificia Camillas, Madrid, Spain, 2013

Value of Lost Load, P. Cramton & J. Lien, University of Maryland, February 14, 2000

Analysis of Benefits: PSE&G's Energy Strong Program, P. Fox—Penner, et al, The Brattle Group, October 7, 2013 on behalf of Public Service Electric and Gas Company, Newark New Jersey

Customer Worth of Supply and Application to Future Regulation of Distribution Systems, G. Strbac & R. Allan, Manchester Centre for Electrical Energy, UMIST, Undated

Electricity Supply Reliability—Estimating the Value of Laod, K.G. Willis & G.D. Garrod, Energy Policy, 1997

Putting a Value of Reliability: Iberdrola USA's Distribution Automation Cost Benefit Analysis, l. Brown, Iberdrola USA, Presented to IEEE Power and Energy Society Annual Meeting, Vancouver, Canada, July 2013

Daily Life in the United States, Kvig, 1920—40

Using Contingent Valuation Method to Explore Willingness to Pay for Electricity and Water Service Attributes, E. Akcura, University of Cambridge, www.electricitypolicy.org.uk, 2008

Estimated Value of Service Reliability for Electric Utility Customers in the United States, M. Sullivan, et al, Sullivan & Co., http:certs.lbl.gov/pdf/lbnl-2132e.pdf

NEMA Electro-industry, The National Electrical Manufacturers Association, May, 2014

Health Care and Social Change in the United States: A mixed system, a mixed blessing, B. Fetter

The improving health of the United States: 1850—1915, Explorations in Economic History—9, 1972

New York in the 1870s, Ephemeral New York, Worldpress, June 12, 2012

Lights and Shadows of New York Life, J. McCabe, 1873

Historical Divorce Rate Statistics, A. M. Jones

Women's Work Never Done, West Virginia Farm Women 1880s—1920s, S. Eagan, West Virginia Archives and History, Volume 49, 1990

A Theory of Human Motivation, A. Maslow, Psychological Review, 1943

Library of Congress, Prints and Photographs Division LC—USZ62-2662

Microgrids and High-Quality Central Grid Alternatives: Challenges and Imperatives Elucidated by Case Studies and Simulation, Daniel Schnitzer, Thesis, Carnegie Mellon University, 2014

United Nations Development Programme. 2006. Human Development Report 2006. New York: United Nations Development Programme.

The Car Disrupted: The 10 most Transformative Ideas, People and Technologies Shaping the Automobile this years, M.D. Paula, Popular Science, October 2014

Built to Crash, Popular Mechanics, October 2014

Smart Cars would Save 420 Million Barrels of Oil over 10 years, W. Jones, IEEE Spectrum (http://spectrum.ieee.org/cars) Sept. 29, 2014

Mercedes Shows Off Self-Driving" Future Truck 2025," P. Ross, IEEE Spectrum (http://spectrum.ieee.org (cars) Sept. 29, 2014

Electrical Safety Foundation International (ESFI), Injury and Fatality Statistics, Arlington, VA 2014 www.esfi.org/index.cfm

Edison vs. Westinghouse: A Shocking Rivalry: The inventors' battle over the delivery of electricity was an epic power play, G. King, Smithsonian.com, Oct. 22, 2011.

The Work of Faraday and Modern Developments in the Application of Electrical Energy, W. Th. Mitkewich, Science at the Crossroads, 1931

George Westinghouse: Gentle Genius, Q.R. Skabec, Algora Publishing, 2007

The History of the Electric Telegraph and Telegraphy, M. Bells, About Money, http://inventors. About.com/od/startinventions, 2014

The Story of Steinmetz, General Electric Company, April 26, 1940

Friend for Life: Robots can already vacuum your house and drive your car. Soon, they will be your companion, A. Piora, Popular Science, November 2005

Appendix A

ELECTRICITY'S PIONEERS

There were several individuals who could be credited with the development of electricity as a wide spread functional system. In order to speculate as to life without it, their collective efforts would have to have failed.

There were several individuals who could be credited with the development of electricity as a wide-spread functional systems. In order to speculate as to live without it, their collective efforts would have to have failed.

Michael Faraday—September 22, 1791 to August 25, 1867

Faraday discovered the electro-magnetic induction of current. He had an intuitive capacity to look into the very nature of things, and arrive at a clear understanding of all that was going on. He recognized that electromagnetic induction enables mankind to transform mechanical work into electrical energy. Faraday's discoveries were the basis of up-to-date electrical engineering, and all the applications of electrical energy. Faraday's scientific work focused on the construction of electromagnetic machinery and other apparatus embodying these ideas in practical use. The fundamental thought that guided Faraday's investigations, and led him to the discovery of electro-magnetic induction, was that between the phenomena of electricity and those of magnetism there must exist a close connection. Given that electricity was flowing through the one, it set up magnetism in the other. What was the converse? Searching from all angles for a solution of this question and continually varying his experiments Faraday was making his way to his aim to "Convert magnetism into electricity." At last, in the autumn of 1831, he solved the problem. He succeeded in generating electric current by means of electro-magnetic induction.

Faraday has stated that always when the conductor is moving across magnetic lines a tendency (electromotive force) develops in this conductor, and electric current is caused if the conductor forms a part of some closed circuit. (Mitkewich, 1931).

James Maxwell Clark—1831—1879

James Maxwell Clark established the principles which define the presence of electromagnetic waves and established the concept that electricity can be generated by moving a coil of wire through a magnetic field. His work is summarized in four equations which define the relationship between electricity and magnetism describing how both current and changing electric fields can give rise to magnetic fields.

Benjamin Franklin—January 17, 1706 to April 17, 1790

Benjamin Franklin was one of the Founding Fathers of the United States. A noted polymath—a person whose expertise spans a significant number of difficult subject areas, Franklin was a leading author, printer, political theorist, politician, postmaster, scientist, civic activist, statesman, and diplomat. As a scientist, he was a major figure in the American Enlightenment and the history of physics for his discoveries and theories regarding electricity. He invented the lightening rod, bifocals, the Franklin stove, and a carriage odometer. He facilitated many civic organizations, including the first fire department and a university. (Source: Eurelectric).

Benjamin Franklin was an early experimenter with electricity. His kite-powered, lightning-enabled exploits were chronicled in his book, "Experiments and Observations on Electricity," which gained public acclaim. What the public did not know is that Ben did not directly endure a lightning charge. Georg Richman, a Swedish scientist living in St. Petersburg, Russia, tried to replicate Ben's work and was instantly electrocuted. (Jones 2004)

Most all of us have watched lightning during a storm and wondered about its power. In order learn more about electricity, Ben conducted the now famous experiment. But his interest in electricity was not just limited to lightning. He received an electricity tube from his friend Peter Collinson and began to play around with it, performing experiments.

Ben suspected that lightning was an electrical current in nature, and he wanted to see if he was right. One way to test his idea would be to see if the lightning would pass through metal. He decided to use a metal key and looked around for a way to get the key up near the lightning. As we all know, he used a child's toy, a kite, to prove that lightning is really a stream of electrified air, known today as plasma. His famous stormy kite flight in June of 1752 led him to develop many of the terms that we still use today when we talk about electricity: battery, conductor, condenser, charge, discharge, uncharged, negative, minus, plus, electric shock, and electrocution.

Ben understood that lightning was very powerful, and he also knew that it was dangerous. That's why he also figured out a way to protect people, buildings, and ships from it. He invented the lightning rod.

Source: www.learn.fi.edu/franklin/scientist/electric.htm

Thomas Edison—February 11, 1847 to October 18, 1931

Edison was an American inventor and businessman. He developed many devices that greatly influenced life around the world, including the phonograph, the motion picture camera, and a long-lasting, practical electric light bulb. Edison is the fourth most prolific inventor in history, holding 1,093 U.S. patents in his name, as well as many patents in the United Kingdom, France, and Germany. (Source: Eurelectric).

After Edison developed the first practical incandescent light bulb in 1879, the rush to build hydroelectric plants to generate DC power in cities across the United States practically guaranteed Edison a fortune in patent royalties. But some increasingly recognized that it was very difficult to transmit over distances without a significant loss of energy. The future of electric distribution, was in AC (alternating current)—where high-voltage energy could be transmitted over long distances using lower current—miles beyond generating plants, allowing a much more efficient delivery system.

Westinghouse Electric began installing its own AC generators around the country, focusing mostly on the less populated areas that Edison's system could not reach. But Westinghouse was also making headway in cities like New Orleans, selling electricity at a loss in order to cut into Edison's business. By 1887, after only a year in the business, Westinghouse had already more than half as many generating stations as Edison. The concern at Edison was palpable, as sales agents around the country were demoralized by Westinghouse's reach into rural and suburban areas. (King, 2011)

Nikola Tesla—July 10, 1856 to January 7, 1943

Tesla was a Serbian-American inventor, electrical engineer, mechanical engineer, physicist, and futurist best known for his contributions to the design of the modern alternating current (AC) electrical supply system. Tesla is also known for his high-voltage, high-frequency power experiments in New York and Colorado Springs which included patented devices and theoretical work used in the invention of radio communication, for his X-ray experiments, and for his ill-fated attempt at intercontinental wireless transmission in his unfinished

Wardenclyffe Tower project. (Source: Eurelectric) After coming to the U.S. from Serbia/Croatia, Tesla worked for Edison solving DC problems and later left Edison when he would not develop and commercialize some of Tesla's own AC inventions. Westinghouse bought Tesla's AC induction motor patent (and others). Tesla then worked for Westinghouse to manufacture and commercialize his AC polyphase electric grid vision. Later in life, Tesla left Westinghouse to develop wireless communication (i.e., the Radio), florescent bulbs, and "wireless electric transmission." This later "dream" never came to fruition.

Nikola Tesla was an eccentric—and unbelievably under-rated—genius known as the 'wild man of electronics', was one of the greatest minds in the history of the human race. If it weren't for this slightly manic genius, today's power systems and many electric end-use devices we enjoy may not have evolved. Tesla invented the alternating-current generator that provides light and electricity, the transformer through which it is sent, and even the high voltage coil of picture tubes. The Tesla Coil is used in radios, television sets, and a wide range of other electronic equipment—invented in 1891, no-one's ever come up with anything better. (Kerry 2014)

Born in Austro-Hungary (now Croatia) in 1856, Tesla constructed his first induction motor in 1883 and immigrated to America in 1884—arriving in New York with only four cents, a pocket full of poems, carefully worked out calculations for a flying machine, and a head full of dreams. Tesla began working with Thomas Edison, but the two men were worlds apart in both their science and cultures and they soon went their separate ways.

George Westinghouse promptly snapped up the patent rights to Tesla's alternating-current motors, dynamos, and transformers. The buy-out triggered a power struggle which eventually saw Edison's direct-current systems relegated to second place, and the DC motors installed in German and Irish trains only a few years before, rendered obsolete.

Tesla went on to set up his own lab. Within a short time, he had preempted Wilhelm Rontgen's discovery of X-rays with his own experimental shadowgraphs; the relays, vacuum tubes, and transistors of future decades with his electric logic circuits; even the wireless radio—the principles of which were described by Tesla in minute detail years before Marconi transmitted his first Morse code message.

Always, Tesla returned to electricity—Tesla loved electricity, and was fascinated with lighting, speaking often of some future world filled with electric light. In a determined effort to prove the safety of his new alternating-current lighting system, Tesla would often light lamps by using his own body as an electrical conductor, to the somewhat muted cheers of an uneasy audience. Turning to studies of resonance, by 1898 Tesla had designed an oscillator that generated half a million volts.

During the Panic of 1907, Tesla ensured the survival of the Westinghouse company by giving up patent payments for a nominal sum. By doing so, he also ensured his own financial ruin. When World War One began in 1914, Tesla lost his payments from European patents. By 1916, he was living in poverty and had filed for bankruptcy to escape a massive tax debt. Soon afterwards, Tesla began to show symptoms of Obsessive-Compulsive Disorder. The symptoms became pronounced very quickly, and what was left of his already 'eccentric' reputation was soon in tatters.

Still, he wouldn't give up. Tesla turned down an attempt by European friends to raise funds and continued to exist on a modest pension from Yugoslavia. At the age of 81, Tesla challenged Einstein's theory of relativity, announcing that he was working on a dynamic theory of gravity that would do away with the calculation of space curvature. The theory was never published, but a similar theory involving gravity waves—developed in the mid-1990s—is used in the study of plasma cosmology (which explains properties of energy and the structure of the universe by studying the electromagnetic effects of plasma).

Even apart from the first AC motor, the radio, the Tesla Coil, vacuum tubes, X-rays, and hydroelectric generators, Tesla had time to develop:

- The loudspeaker
- Fluorescent lights
- Radar
- The rotary engine
- Microwaves
- The basis for diathermy (deep heating tissues through the use of high-frequency electrical current), and
- An 'automatic mechanism controlled through a simple tuned circuit'—remote radio control.

Source: Kerry Redshaw, Brisbane Australia, Nicola Tesla, www.kerryr.net/pioneers/tesla.htm

George Westinghouse—October 6, 1846 to March 12, 1914

George Westinghouse, had the ability to understand the significance of new ideas and technology could also redesign them to enable commercial manufacturing and ease of maintenance. Westinghouse bought numerous patents from Tesla in his efforts to commercialize AC electricity. When Tesla needed money to live on, in his later years, Westinghouse paid for Tesla's room and board at the Waldorf Astoria Hotel in Manhattan.

Lord Kelvin referred to Westinghouse' life as "The electric development we know today would have long halted without his daring and resourcefulness." Actually Westinghouse' interest in electricity started with applying electricity to railroad brakes and signaling. He had a knack for converting abstract scientific ideas into practical electrical engineering. After attending the 1876 Philadelphia Centennial Exposition, where various electric generator were on display pictured as driven by Westinghouse Steam engines. These dynamos represented a large market for Westinghouse and Machine Company. Westinghouse approached his business interest in electricity with a search of everything written on the subject.

By 1879 Westinghouse was experimenting with lighting systems. His main interest was the use of arc lighting in his own factories. His interest in AC systems evolved slowly at first. Westinghouse was convinced that Ac systems were superior, in part because they were conceptually similar to Westinghouse's gas system in which high pressure pipelines carried gas longer distances dropping pressure in subsequent distribution systems and the system approached end-use consumers. (Skarsse, 2007)

Charles Steinmetz—April 9, 1865—October 26, 1923

Charles Proteus Steinmetz was a mathematician, who became an electrical engineer/inventor of complex algebra applied to AC machines and networks. In 1894 the General Electric Company transferred its operations to Schenectady, N.Y., and Steinmetz was made head of the calculating department. He at once began to indoctrinate the engineers with his complex algebraic method of calculating alternating-current circuits. During the next 20 years he prepared a series of masterful papers and volumes which reduced the theory of alternating current to order enabling performance prediction before construction.

Steinmetz's first forays into electrical engineering involved investigations into the losses within magnets used in AC motors. This led to the development of what is referred to as the law of Steinmetz, called the "law of hysteresis loss." His career brought him to the Calculating Department of the General Electric Company where he developed most of the principles and design basics for alternating current machines. These principles are today taught in all engineering schools and are used in practically every application of alternating current. (General Electric, 1940).

Index